Electronic Projects 4

Test Gear Projects

Terry Dixon

First published 1980 by
THE MACMILLAN PRESS LTD
London and Basingstoke
Associated companies in Delhi Dublin
Hong Kong Johannesburg Lagos Melbourne
New York Singapore and Tokyo

Typeset in 10/11 Univers by
STYLESET LIMITED · *Salisbury, Wilts.*
and printed in Great Britain by A. Wheaton & Co. Ltd., Exeter

British Library Cataloguing in Publication Data

Electronic projects.
4 : Test gear projects
1. Electronic apparatus and appliances – Amateurs' manuals
I. Dixon, Terry
621.381 TK9965

ISBN 0–333–26396–0

Contents

Preface

The advent of cheap, widely available transistors and integrated circuits has made electronics a very interesting and reasonably economical hobby. Once bitten by the bug, you will soon be faced with a number of problems. What if you can only get hold of a design that does not quite do what you want it to? What do you do about a project that, once built, does not work? Troubles like these are much easier to solve if you are equipped with an array of reliable test gear. In this book, I have assembled a variety of projects enabling you to measure what you are doing rather than your having to guess!

Assuming that electronics construction is not just a fleeting passion, there are several instruments that will be almost essential in a well-equipped workshop, and I have suggested designs for most of these, although there are one or two important pieces of equipment that I have decided not to include. Digital instruments are coming down in price all the time and have now reached a level where it may be cheaper to buy rather than build. An oscilloscope is an asset to any workshop but I feel that the problems involved in building one of adequate performance would stretch the most experienced home constructor. I think it may be a better idea to buy a modest oscilloscope and later extend its performance with the accessories that I have included.

If you are considering buying test gear, caution should be your watchword. High-grade apparatus, by well-known makers has always commanded a good price and you must expect to pay for quality. Although you may find a bargain in the form of good-quality second-hand equipment, do make sure that it fits your requirements and, of course, that it is working properly.

In designing the various projects, I have endeavoured to keep them as simple as possible — conducive with adequate performance — avoiding the use of any specially wound coils and unusual components.

Almost any electronic project that you can think of needs a power supply, and in chapter 1 I have included a selection of power packs which should meet most of your requirements. The first one is a small series stabiliser, which you can use to regulate a battery supply or can be fitted directly on to your circuit board. The second supply is particularly suitable for use with your CMOS or TTL projects but you can use it to supply any circuit needing between 5 and 15 V with up to about 200 mA. Most op-amps require a balanced supply and this is provided by the simple dual polarity power pack, but if you are building projects using mixed circuitry then the balanced supply from one winding may suit your needs — this gives equal positive and negative output rails when a heavier current is being drawn from the positive one. The *200 V from a PP3* circuit enables you to operate high-voltage low-current projects from a standard 9 V battery. You can find similar circuits in many electronic flashguns and in the fluorescent light fittings used by campers. This design uses a voltage doubler which is a very useful technique for obtaining high voltages. I have shown two more doubler circuits, one using a twin electrolytic capacitor of the type which may easily be salvaged from many scrap TV chassis. You may find either of my blown fuse indicators or the crowbar protection circuit useful additions to any of your workshop power supplies. The crowbar technique is used in many colour TV power supplies to protect the expensive specialist integrated circuits used.

Chapter 2 deals with sources of signal and I think I have provided designs to cover most of the testing and servicing jobs that you may be carrying out at home. The signal injector and the a.f. generator are simple, general purpose instruments. You can nevertheless locate the faulty stage in a superhet radio receiver using only the signal injector. The logic probe will indicate high or low logic levels in TTL circuits and you will find that the crystal reference oscillator and the waveform generator give an excellent performance for such simple designs. Both of these circuits demonstrate the advantages of integrated circuits, which are less expensive, more compact and much easier to use than their discrete component equivalents. If you are interested in audio electronics, I think you will find the white noise generator and the stereo balance quite useful.

Chapter 3 deals with test meters and their accessories. Multimeters are precision instruments and it is not easy or cheap to construct one at home. My dc multimeter is based on a 0 to 100 μA moving coil movement but I have included enough design information for you to adapt the circuit for any movement that you have to hand. In this context I think you will find my circuit for measuring meter internal resistance very helpful. The most expensive component in any multimeter is likely to be the movement itself and I think there is a strong case for buying one of the many excellent, reasonably priced, ready built instruments available – particularly if you need an ac meter. My high-impedance voltmeter uses a similar 0 to 100 μA moving coil movement but the input impedance of the meter is raised to about 10 MΩ by means of an op-amp. You can use the r.f. probe in conjunction with this meter to indicate the level of r.f. voltage in radio circuits.

It is good practice to check all your electronic components before you use them, particularly if you intend to use unmarked ones or devices salvaged from old circuit boards. In chapter 4 I have included a wide variety of component checkers. A technique rather than a specialised circuit is needed to estimate the value of an unknown capacitor but, if you have an oscilloscope and an a.f. generator, you can get a quick idea of the approximate value of an unknown capacitor. The simple capacitance bridge gives a more precise measurement of smaller capacitors and the electrolytic capacitor tester can be used with high-capacitance low-voltage electrolytic capacitors. The diode tester is a simple go/no-go indicator and you will find it very useful for sorting out batches of unmarked, untested components. The transistor tester enables you to sort out faulty components and to measure the static gain of the good ones.

In chapter 6 I have suggested a few accessories to extend the usefulness of your oscilloscope. The calibrator enables you to check the calibration of your oscilloscope from time to time. If you have a single trace oscilloscope, you will find the dual trace adaptor very useful. Among other things this enables you to compare two signals and make accurate assessments of comparative voltage, frequency and phase shift. The sawtooth generator can be used as an external timebase which is particularly useful for those awkward jobs where it is difficult to lock the display.

The circuits set out in chapter 6 are not really test instruments, but are the aids to servicing that no workshop can do without. I think you will find them all useful at some time or other, particularly if you can develop the ability to adapt and improvise where the suggested circuits do not meet your particular requirements. Whatever branch of electronics you are interested in, I am sure that some of these designs will help you with your project work and I hope that they will result in you getting even more pleasure out of our hobby.

TERRY DIXON

Electronic Projects 1 – Cost-effective Projects Around the Home
John Watson

Electronic Projects 2 – Projects for the Car and Garage
Graham Bishop

Electronic Projects 3 – Audio Circuits and Projects
Graham Bishop

Electronic Projects 4 – Test Gear Projects
Terry Dixon

Publisher's Note

While every effort has been made to ensure the accuracy of the projects and circuits in this book, neither the Publishers nor the author accept liability for any injury or loss resulting from the construction of any of the designs published herein.

The Publishers will, however, be pleased to hear from readers who have corrections to the text or genuine queries, and will refer any such queries to the author.

1

Power Supplies

Double your Voltage with a Twin Electrolytic

If you need a high-voltage low-current supply, try a voltage multiplier circuit. A straightforward step-up transformer would give you a high voltage, but a shock received from this type of supply could well be lethal. The relatively high output impedances of most multiplier circuits render them much safer to use, although high-voltage circuits must always be treated with respect.

Circuit

The basic suggested circuit is shown in figure 1.1. During the first half-cycle, D1 charges C_1 to the peak value of the input and during the second half-cycle D2 charges C_2 to the peak value of the input. Since C_1 and C_2 are connected in series, the unloaded output is about twice the peak voltage of the ac input signal from the transformer. The capacitors and diodes should have a minimum rating of the peak value of the input voltage – that is about 1.5 times the r.m.s. value. For example, if you use a 1:1 transformer connected to a 240 V mains supply, the peak secondary voltage will be about 340 V. To ensure good reliability, you should use components rated at 400 V minimum.

The output regulation is largely dependent on the capacitance of

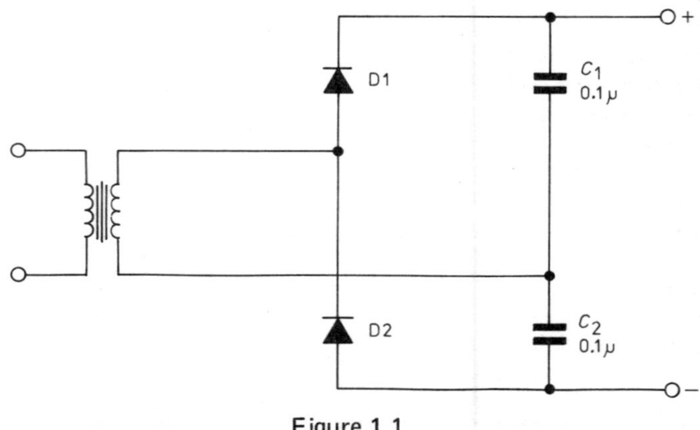

Figure 1.1

C_1 and C_2 and for most high-voltage applications polyester or polycarbonate types are most suitable, but for lower voltages you can get better regulation with electrolytic types.

Figure 1.2 shows how the basic circuit can be modified to use a twin electrolytic capacitor. During the positive half-cycle, C_2 charges up via D1 and during the negative half-cycle C_1 charges up via D2. Regulation is much better than for the circuit shown in figure 1.1, but remember that using larger capacitors means that you can get a much nastier shock from it!

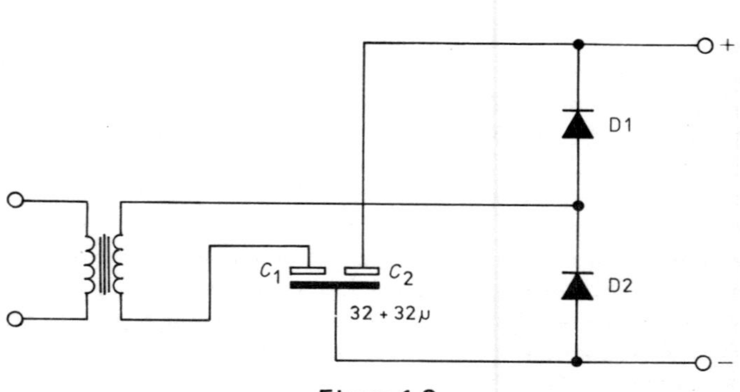

Figure 1.2

Simple Regulator

You may well have come across those extremely useful timer ICs, the 555 and the 556. In some applications, the frequency stability of these devices is greatly improved if the chip is provided with a stabilised supply.

My suggested circuit will supply several ICs and is compact enough to fit directly on to a circuit board. Alternatively, it can be used to supply a 'test bed' when you are experimenting with these chips.

Circuit

The circuit is shown in figure 1.3. TR1 (which is a BC107) is a conventional series regulator. R_1 and Z1 form a shunt stabiliser, holding the base of TR1 at a constant 13 V. The circuit is supplied by two 9 V batteries in series and I have found that sufficient decoupling is provided by the small capacitors C_1 and C_2. With the type of Zener diode specified, the regulator gives about 12 V output. It will provide up to 50 mA output current, but TR1 does not require a heatsink unless it is going to be run flat out for long periods of time. In this case, fit one of the small, push-on heatsinks.

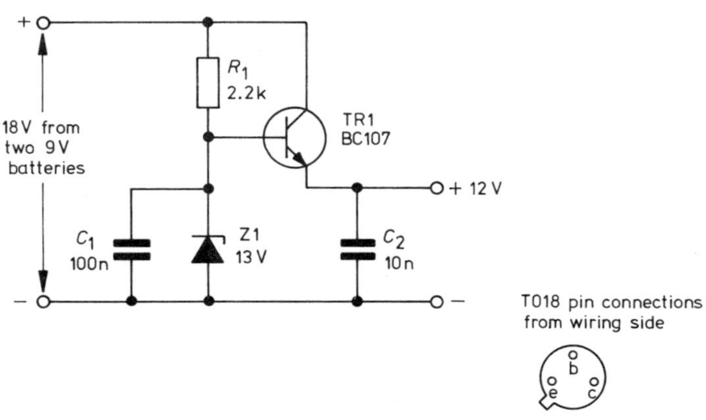

Figure 1.3

Regulated Power Supply for CMOS and 74 Series ICs

The 74 series TTL IC requires a 5 V power supply. The 4000 series CMOS IC has a recommended maximum supply voltage of 12 V (AE suffix types) or 18 V (BE suffix types).

The circuit here will provide either a fixed 5 V regulated supply or one which can be varied from about 6 to 15 V and uses a 7805 regulator and a 741C op-amp. The complete circuit is shown in figure 1.4.

Circuit

The transformer T1 is a 9—0—9 V type with its secondary rated at 250 mA at least. The bridge rectifier (D1) is a WOO5 which is rated 50 V at 1 A. The value of C_1 is not critical but you should not use a much larger one than this, since the switch-on surge current may be too heavy for D1. C_2 is mounted close to the regulator to ensure that it remains stable.

With S2 in the 'fixed' position, only half the secondary of T1 is applied to D1. The common terminal of the regulator (REG) is returned to chassis. The output is a fixed +5 V regulated and the 741C is inoperative. C_3 is fitted to improve the transient response of the regulator.

With S2 in the variable position, the entire secondary of T1 is applied to D1. The common terminal of the regulator (REG) is connected to chassis via a 1 kΩ resistor. The output voltage is sampled by the chain of resistors R_2, VR_1 and R_3. The wiper of VR_1 is taken to the non-inverting input of the 741C op-amp. The amplifier has unity gain and its output is connected to the regulator (REG common terminal). Adjusting VR1 gives an output voltage of about 6 to 15 V.

With the recommended transformer, output current is limited to about 200 mA.

Construction

I built my power supply into a small aluminium box of about 100 mm x 60 mm and 50 mm deep and fitted most of the electronics on to a piece of Veroboard, using the layout shown in figure 1.5.

When the supply is used in variable mode, with the output voltage set to the lowest level, the regulator (REG1) will have to dissipate about 4 W (assuming 200 mA load current). The 7805 (REG1) must be mounted on a heatsink. Mine was in the TO—220 plastic package so I fitted it — using an insulating fixing kit and a little silicone grease — to a finned heatsink rated at about 20° C/W. This whole assembly I bolted to the side of the aluminium box, which provides cool reliable running under all conditions. If you use a device in the TO—3 type package, sufficient cooling will be achieved if you fit it to the outside of the box — once again using an insulating fixing kit and silicone grease.

input
com
output
7805
regulator
L
N
mains
input
FS1
S1a
S1b
A
B
T1
ii
i
w
S2a
D1
W005
in
reg 1
7805
out
com
w
S2b
i
ii
C1
2200μ
40V
C2
330n
R1
1k
C3
100n
IC1
741
7
6
4
2
3
R2
3.3k
VR1
5k
lin
R3
3.3k

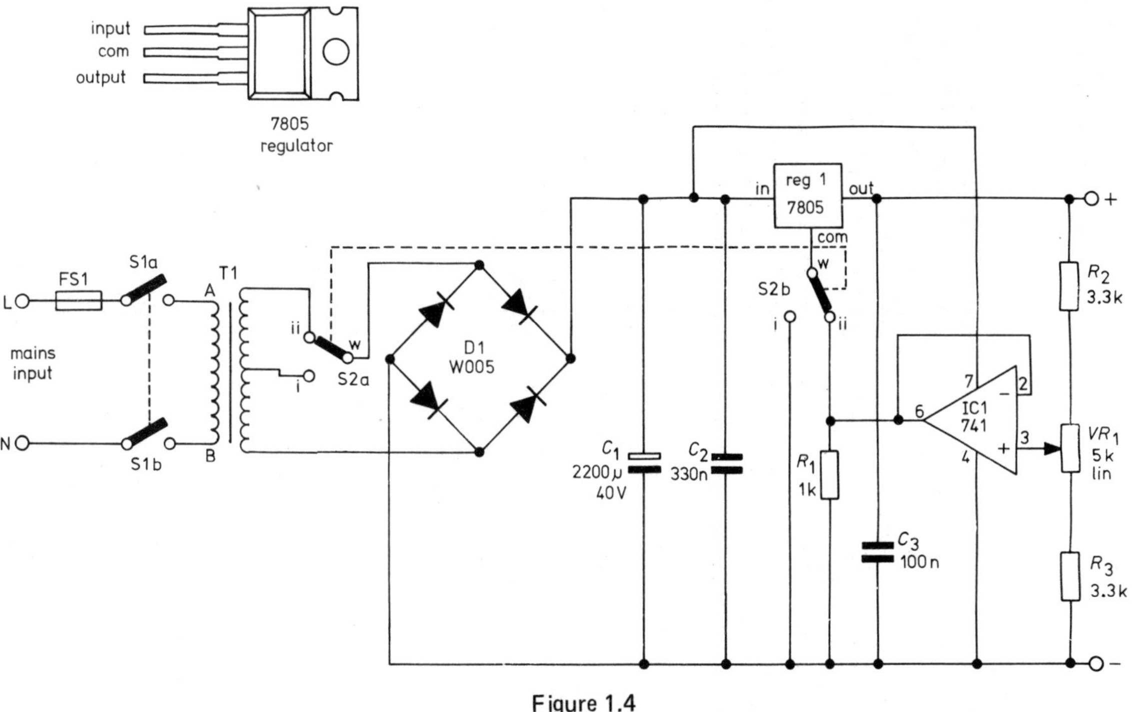

Figure 1.4

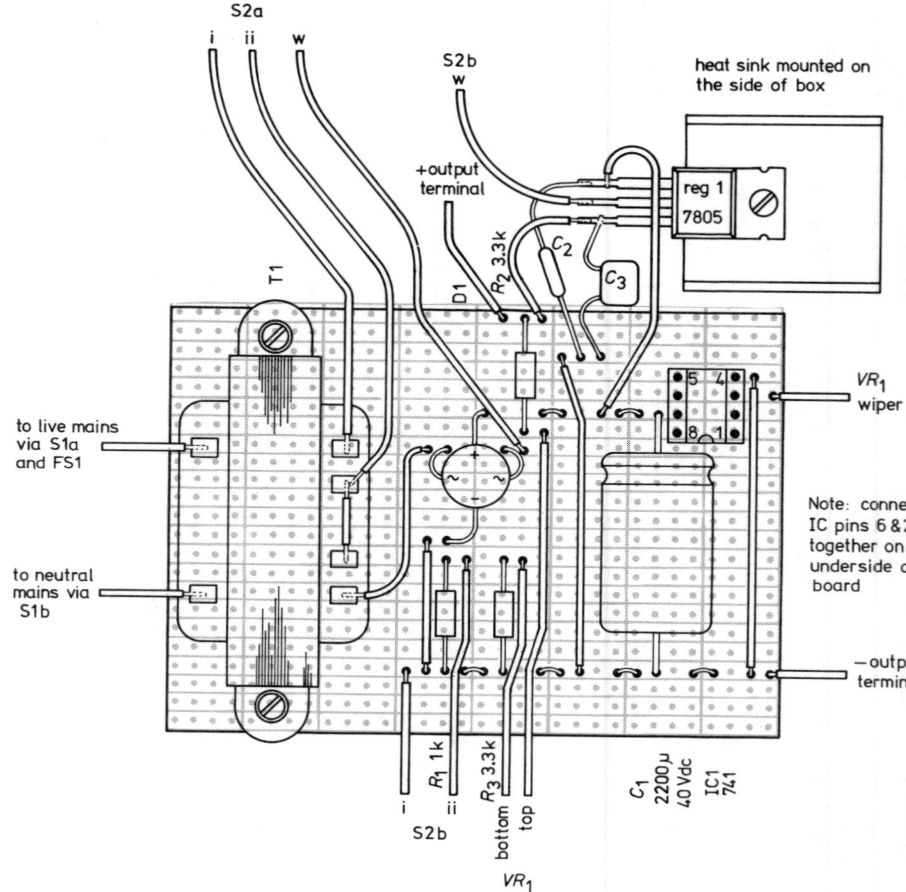

Figure 1.5

I mounted VR_1, S1, S2 and FS1 (using a screw-type fuse holder) on the front panel of the box together with the output terminals. If you wish you can also fit a neon mains indicator (across AB in figure 1.4) to the front panel. This should have a built-in limiting resistor. If not, you must fit a $\frac{1}{2}$ W resistor of about 330 kΩ in series with the neon. Be careful that none of the mains side wiring gets anywhere near the sides of the box.

Make sure that the 741C is mounted round the right way. (Use a socket to mount it on the board, or if not, use a small bulldog clip – shorting all the pins together – as a heatsink when soldering in the IC.) Leave the leads on the WOO5 (D1) full length and insulate them with short lengths of PVC sleeving. Don't forget that C_1 is polarised and must be fitted the right way round. Generally, the negative end is connected to the capacitor casing, but there are so many types that, if you are not sure, ask about the polarity when you buy the capacitor. I didn't bother with a mains input socket, and simply took a mains lead out of the side of the box using a small Neoprene grommet.

Testing and Fault-finding

When you have completed all the wiring and checked it over, switch S2 to the fixed position – don't forget to fit a fuse to the front panel fuse holder! Connect the supply to the mains and switch on. Using a high-resistance voltmeter, you should be able to read 5 V across the output terminals. Fit a 25 Ω, 1 W resistor across the output. The voltmeter should remain at 5 V.

Switch S2 to the variable position with the voltmeter still in place – switch the voltmeter to a higher range first, if necessary. Adjusting VR_1 should now give an output voltage variable between about 6 V and 15 V.

If the main fuse blows when you switch on, this can be caused by:

(a) shorted turns on the primary or secondary of T1;
(b) one diode in the bridge rectifier D1 short circuit;
(c) the reservoir capacitor (C_1) short circuit;
(d) the decoupling capacitor (C_2) short circuit;
(e) any of the mains or secondary winding wiring shorting to earth.

These must be investigated and any faults found rectified before switching on again. *Disconnect the circuit from the mains and make resistance checks of these components.*

If there is no output at all when you switch on and REG feels warm, then C_3 is probably short circuit.

If the output voltage is wrong, then this may indicate that C1 or one diode in D1 are open circuit. However, these faults are masked by the very efficient regulation of the 7805 and unless the average voltage at its input falls to less than about 10 V, the circuit will appear to operate normally. A high output voltage that is not influenced by adjusting VR_1 indicates a short circuit 7805 (very unlikely) or a faulty 741C. Try changing the 741C, remembering that it must be fitted the right way round.

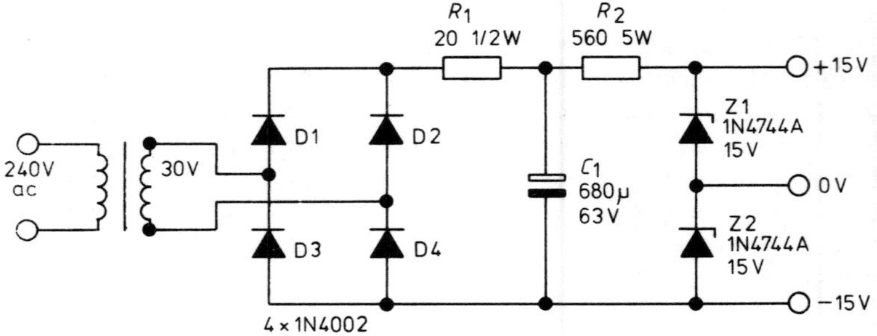

Figure 1.6

Simple Dual Polarity Power Supply

Although many op-amps can be run from a single sided power pack, some circuit designs demand a balanced supply. You can easily build this circuit, as shown in figure 1.6, from readily available components. The output is regulated at ±15 V up to a maximum load current of about 20 mA and so it is capable of driving several op-amps.

A Balanced Supply from One Winding

This circuit, shown in figure 1.7, enables you to drive logic circuitry — requiring a single sided supply — and op-amps — requiring a balanced supply — at the same time. In order to provide equal voltages on both + and — rails, the + one must be more heavily loaded. If you are driving op-amps only, load the rail with a small indicator lamp as shown. In this case R_1 is fitted to limit the lamp current.

200 V from a PP3!

As everyone knows, the normal output voltage from a PP3 is about 9 V. However, the circuit suggested here uses a PP3 — or any other 9 V battery — to drive an oscillator, which, via a step-up transformer and a voltage doubler, provides a supply of about 200 V.

Circuit

The circuit is shown in figure 1.8. A small silicon transistor is con-

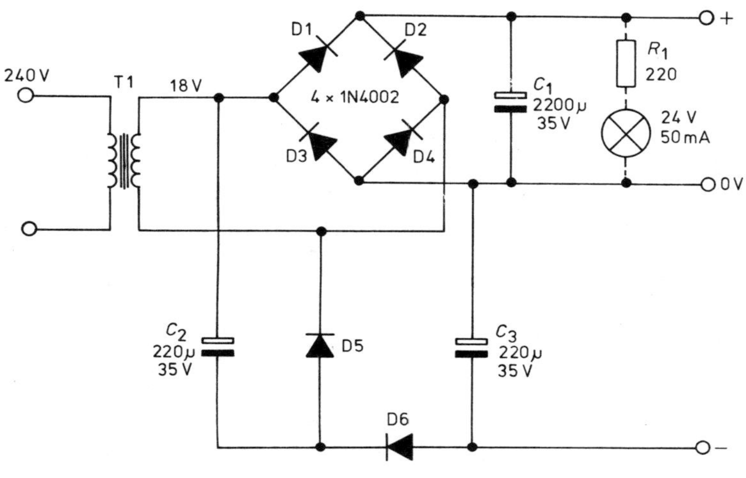

Figure 1.7

nected as a Hartley oscillator using a 250 V to 9 + 9 V mains transformer. Resistor R_1 provides 'starting bias' for the transistor and capacitor C_1 decouples the battery supply. The 'mains' side of the transformer is connected to C_4, C_5, D1 and D2, which together form a voltage doubler. It is important that you use the high-voltage components that I have specified. The 200 V supply is at high impedance and will only provide a few milliamps before the volts start to collapse. Although this high voltage should be treated with respect, it is not dangerous unless you are very sensitive to electric shock. Please don't use my circuit to find out!

Construction

A suitable Veroboard layout is shown in figure 1.9. The transformer that I used was a small one and I bolted it directly to the board. Keep the diode leads full length to prevent heat damage when soldering. The unit is a little too big for the normal 25 mm switch box so I found a plastics box (with a metal front panel) about 95 mm x 160 mm x 50 mm deep. The battery can be fitted in this, too, provided that it is firmly located and kept away from the high-voltage side. I used a pair of screw terminals — one red and one black — which would also accept 4 mm plugs for the high voltage output.

+9V
+200V approx.
C5 470n 630V
–
D2 BY127
a
k
D1 BY127
k
a
C4 470n 630V
4
5
1
2
3
T1
TR1 BC107
C3 47n
C2 47n
R1 10k
C1 100µ
b
e
c
BC107 from wiring side
k
a
BY127

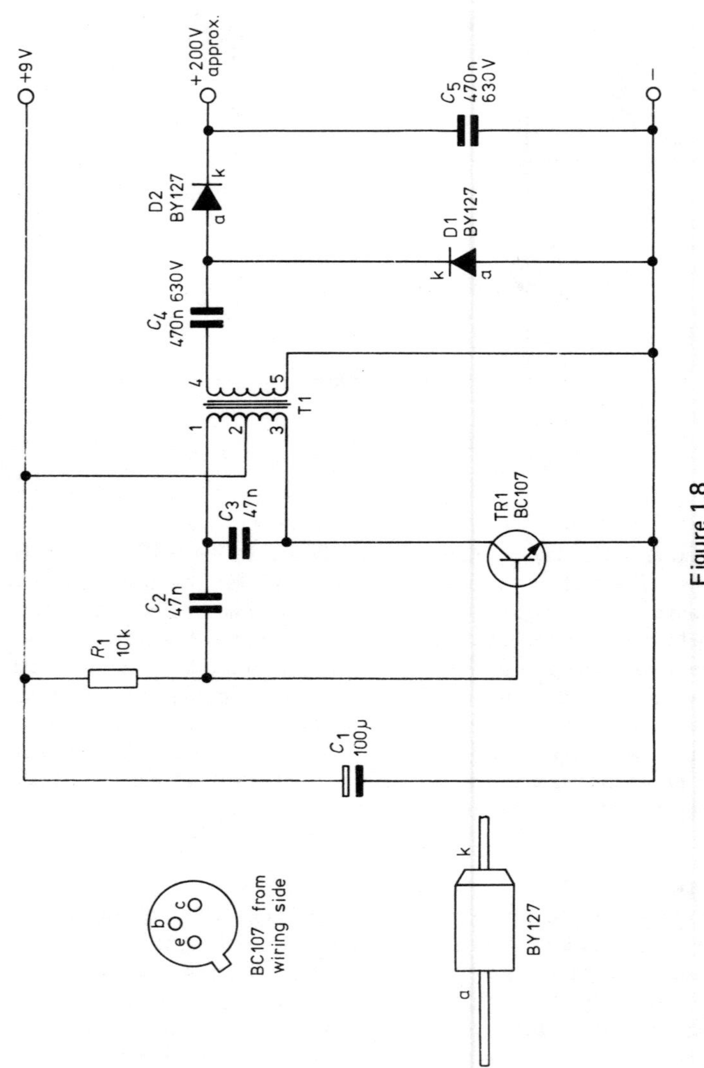

Figure 1.8

+200 V
common
C_5 470n 630V
D2 BY127
D1 BY127
C_4 470n 630V
T1
TR1 BC107
C_2 47n
C_3 47n
+9V
R_1 10k
C_1 100µ

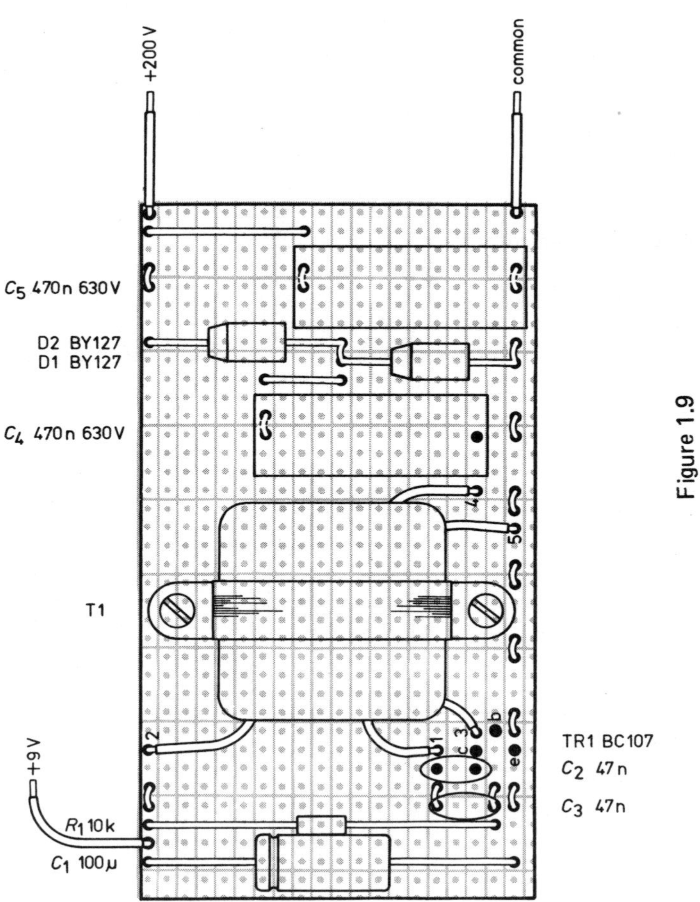

Figure 1.9

Testing and Fault-finding

Connect a high impedance voltmeter to the high-voltage output terminals and fit a known good battery or dc power supply to the circuit. You should get an output of about 200 V dc. If not, then check whether the transistor is oscillating — preferably by observing the collector with an oscilloscope. If not, then try reversing leads 1 and 2 (see figure 1.8) of the transformer, having disconnected the battery first! If this does not produce the desired result and you are sure that your wiring is correct — are C_1, TR1, D1 and D2 the right way round? Try changing the components in the order semiconductors, capacitors, resistors, checking again after each change.

Blown Fuse Indicator I

Referring to figure 1.10, when the fuse blows, the full mains voltage appears across the mains indicator neon. The circuit assumes that the neon has a built-in limiting resistor. If the neon that you use does not, then you must fit a resistor of about 220 kΩ $\frac{1}{4}$ W in series with it.

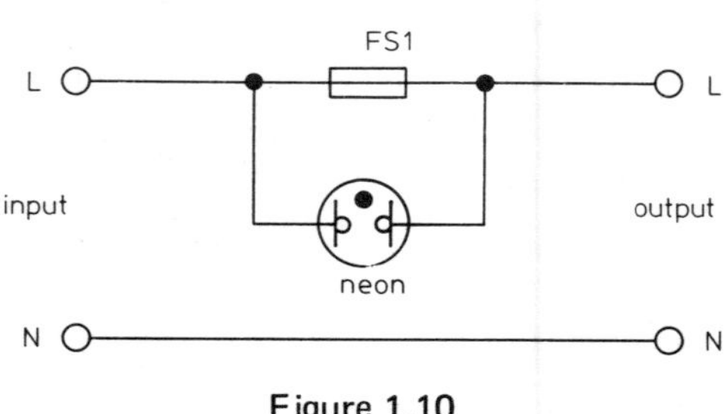

Figure 1.10

Blown Fuse Indicator II

The circuit is shown in figure 1.11. When the fuse FS1 is intact, the neon merely indicates that the mains is on — in the normal way. When the fuse blows, R_1, C_1 and the neon act as a relaxation oscillator. The neon will flash on and off, indicating that the fuse has blown.

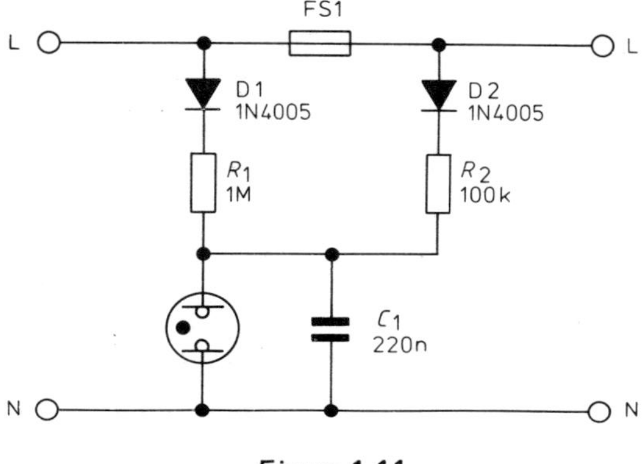

Figure 1.11

Power Supply Protection with a Crowbar

There are some situations where a power supply failure can cost you a lot of money and wasted time. TTL circuitry requires a 5 V regulated supply, which can be provided using a simple series regulator IC. These are well protected against overload but can, and sometimes do, fail under experimental conditions. If the regulator goes short circuit, your TTL circuits will be subjected to the full unregulated dc, which is likely to be in excess of the maximum safe voltage for TTL (usually 7.5 V). A fuse will not protect your circuits – they will blow long before the fuse. The method I suggest you use is called 'crowbar protection' and is to be found in many modern colour TV power supplies.

Circuit

The circuit is shown in figure 1.12. If the voltage at the output terminals rises above about 6.2 V, Z1 conducts charging C_1. This voltage will fire the silicon-controlled rectifier (SCR 1), which quickly shorts – or puts a 'crowbar' – across the supply rails, blowing the fuse (FSI). With all the component types and values that I have specified, the circuit operates very quickly indeed. The SCR clamps the supply rails together long before (in electronics terms a few milliseconds is a long time!) the critical 7.5 V is reached and the fuse (FS1) can blow in its own good time.

I chose a T1C 106A silicon-controlled rectifier because it can handle an anode current of up to 4 A at 100 V, but has a very sensi-

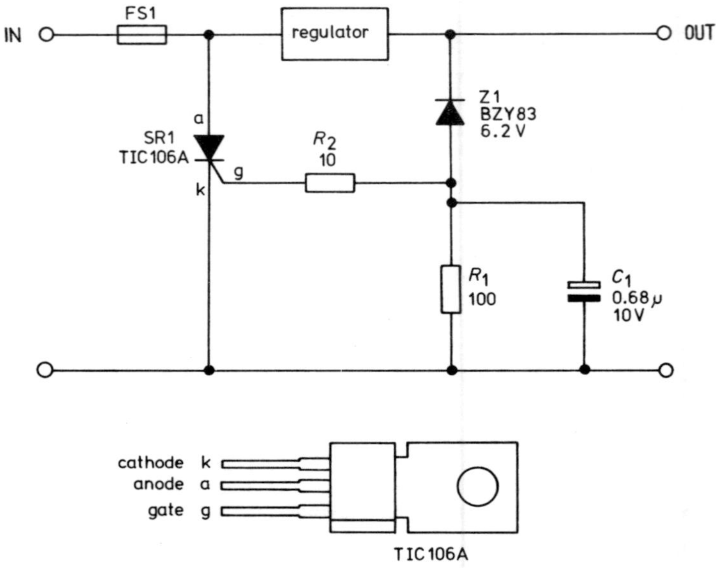

Figure 1.12

tive gate – requiring only about 200 μA to fire it. The SCR is only on for a very short time and thus does not need a heatsink. If you use another type, it is not likely to display the same characteristics. Most SCRs with a heavy anode current rating have insensitive gates. SCRs with a lower anode current rating will need a heatsink. You may find it is possible to bolt such a device to the power supply casing – using an insulating mounting kit, of course.

The rating of FS1 depends on the power supply you are using the circuit with. Assuming a maximum power supply output current of about 200 to 300 mA, you can fit a 500 mA fuse (you must fit it after any large electrolytic smoothing capacitors because these may blow the fuse with their switch on surge). I used a 20 mm cartridge type with a screw-type holder fitted to the power supply front panel.

The voltage at which the SCR switches on is largely dependent on the value of Z1. You can experiment with different values of Z1 to protect other types of circuit. The 4000 series CMOS ICs have supply voltage limits of 15 V (AE suffix) or 20 V (BE suffix) and crowbar protection can be used for these too.

2

Generators and Signal Injectors

A Two-transistor Signal Injector

If you are about to embark on a complete realignment of your communications receiver, then you will need a proper signal generator. However, for simpler routine servicing, it is useful to also have a small, self-contained signal injector to fall back on.

This circuit is inexpensive to build and avoids the use of complex switching or tuning coils. The output waveform is rich in harmonics so that you can use the injector for checking audio stages, or any stages of a medium wave/long wave radio right back to the aerial input.

Circuit

There are several problems to be dealt with when constructing your own signal generator. Most designs for r.f. generators are based on *L—C* oscillators often requiring several switched ranges, the winding of tuning coils and the alignment of the completed generator against some reliable standard. Any of these difficulties can be overcome with a little care and trouble, but this design avoids them altogether by making use of the fact that a squarewave is made up of a fundamental frequency plus a great number of harmonics.

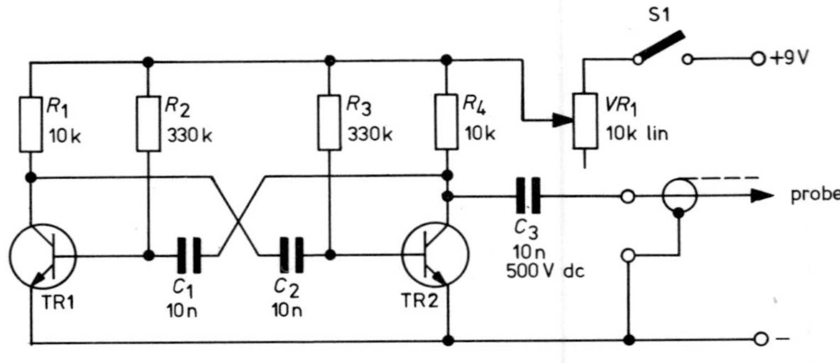

e b c

TR1 & TR2 pin connections
from wiring side

Figure 2.1

The circuit shown in figure 2.1 is basically a free-running (astable) multivibrator. The supply voltage is 9 V using a PP3 battery fed to the circuit through a variable resistor, which acts as a signal output level control, and a push button, which makes sure that you do not accidentally leave it switched on. The battery should last for ages because only about 1 mA is drawn from it even at maximum output.

Any pair of a.f. transistors will work well – I used a couple of BC107s. You can use *pnp* devices if you like but remember to reverse the power supply polarity. The frequency-determining components are C_1, C_2, R_2 and R_3 providing a fundamental a.f. output of about 500 Hz. C_3 should be a component with a working voltage of about 500 V dc allowing the injector to be used with valve equipment. If you never work on valve equipment you might like to substitute a smaller component for C_3.

Construction

The wiring and general layout of the circuit is not critical. I fitted all the components except S1 and VR_1 to a very small piece of Veroboard using the layout shown in figure 2.2. The only precaution you need to take is to make sure that you don't cut the transistor leads off too short – keep them at least 6 mm long. Although modern silicon transistors are much tougher than the older devices used to be, try not to heat them up too much. If in doubt, use a small croc-clip as a heatsink attaching it to each lead while you are soldering.

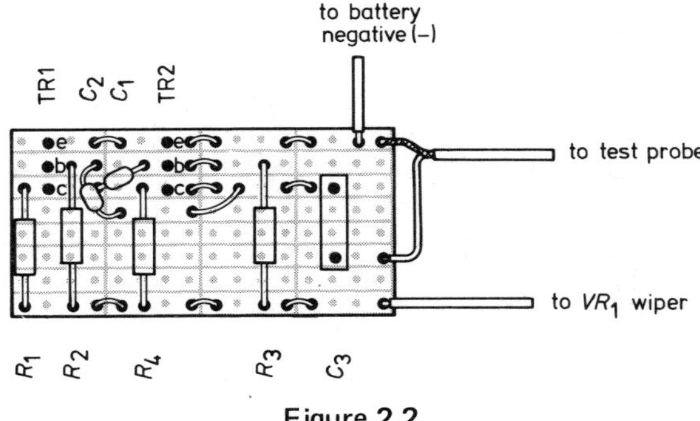

Figure 2.2

I fitted the circuit board, the variable resistor, the battery and the push button switch into a small plastics box. I then took the output, via a flying screened lead with the screen connected to a croc-clip and the centre conductor, to a test probe. If you are more adventurous, you may like to experiment with a pen-torch-type housing with a homemade probe built into the end of it (using a small brass nail, for example). It is always a good idea, when project building, to try to use any bits and pieces that you have to hand.

Fault-finding

When testing any prototype or experimental circuit, I always like to supply it from a stabilised, short circuit protected power supply. In this way you avoid looking for a fault in the circuit when, in fact, you are using a flat battery. The short circuit protection will save the power supply if the prototype draws too much current.

Perhaps the best way to test your circuit is to connect it to an oscilloscope. If you don't have one, you can use an ordinary transistor portable radio. Switch the radio on, setting the tuning dial off station, and the volume control at maximum. Switch on the injector and turn VR_1 to maximum. Hold the test probe near the ferrite rod aerial and you should hear a noisy whistle from the loudspeaker. If not, check your wiring carefully – are you sure that the transistors are the right way round? Has a blob of solder bridged between the two conductors on the circuit board?

The circuit should draw about 1 mA from the supply on full output. If it is drawing much more (or much less) you have probably used faulty transistors. Replace each in turn, and try again. If it still doesn't work, replace the capacitors, and then the resistors, one by one.

A Logic Probe for TTL

For a long time now the multivibrator signal injector probe has been a useful tool for checking analogue circuits. With the increasing availability of cheap logic ICs, the digital logic probe is equally valuable for use with digital circuits.

My suggested circuit indicates logic level '0' and logic level '1' using light emitting diodes (LEDs). I have designed it for use with TTL circuitry – about the most popular logic 'family' at the moment.

Circuit

The circuit diagram is shown in figure 2.3. The IC is a 7404 Hex Inverter and TR1 is any small silicon *npn* device. If a logic '1' level (anything greater than 2.5 V) is applied to the probe tip, TR1 is biased on via R_1. The positive level at TR1 emitter is inverted by one section of the 7404 and this will illuminate D2, the '1' level LED. The small resistor, R_4, limits the current flowing through D2.

If a '0' logic level (anything less than 0.5 V) is applied to the probe tip, TR1 does not switch on. The '0' is inverted twice by two sections of the 7404 illuminating D1, and the '0' level LED R_3 performs a current limiting role for D1.

The power for the probe is picked up from the test circuit itself. Since only three sections of the 7404 are used, you may be able to use a scrap circuit which has a fault on one of its inverters.

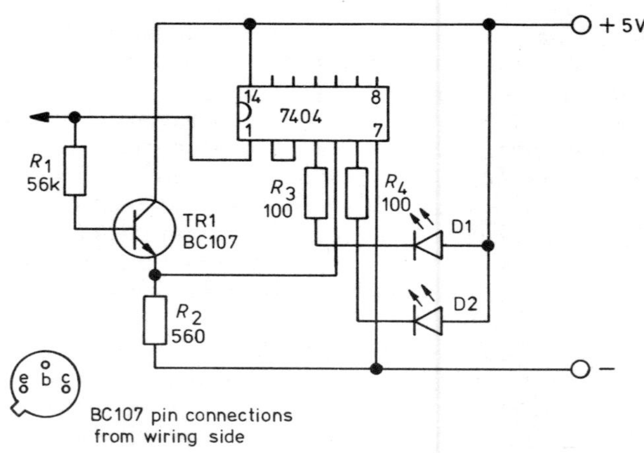

Figure 2.3

Construction

I built most of the components on a small piece of Veroboard, using the layout shown in figure 2.4. I used a 14-pin DIL socket for the 7404 because this makes servicing so much easier. I fitted the board into a small plastics tube. At one end of the tube I fitted a small screw — to use as a probe tip. I fitted the LEDs to the other end and I brought a 600 mm length of single screened lead out of the side of the tube for picking up the supply. I fitted a couple of croc-clips for this purpose.

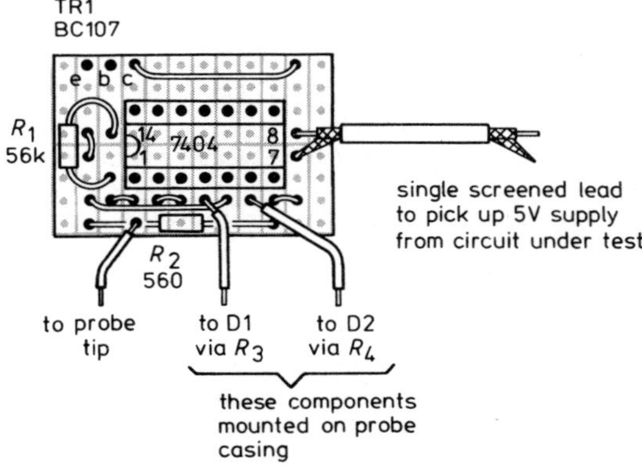

Figure 2.4

Testing and Fault-finding

Connect the power leads to a low-voltage dc power supply set at 5 V. Touching the probe tip on the 5 V terminal should illuminate the '1' LED and touching the probe tip on the 0 V terminal should light the '0' LED.

If the probe fails to function correctly, check you wiring again — just to make sure. If all is correct and you have used a socket for the 7404, unplug it and try another one. You can check D1 and D2 by touching the 5 V on each anode and the 0 V on the left-hand end (see figure 2.4) of R_3 and R_4 in turn. If D1 operates but D2 does not, then you should suspect a fault in TR1. Swap it for another one and try again.

a.f. Signal Generator

A good a.f. signal generator should be high on any constructor's short-list. The design is very easy to build, but is capable of producing a very low distortion sinewave output from about 20 Hz to about 20 kHz over three switched ranges. The generator has an output voltage continuously variable up to a maximum of 4 V peak to peak and is powered by a single 9 V battery such as a PP3.

Circuit

The basic schematic diagram is shown in figure 2.5. The amplifier, R_A/C_A and R_B/C_B form a Wein bridge oscillator. The oscillator frequency (f) is given by $f = 1/2\pi CR$. At this frequency alone the bridge provides zero phase shift but an attenuation of about 3. If it is arranged for the amplifier to have zero phase shift and a gain of 3 then the circuit will oscillate.

The circuit shown in figure 2.6 is designed around a 741C IC op-amp. I decided to use an IC rather than discrete components since it is cheaper, simplifies construction and has the added advantage of built in output short circuit protection. The 741C normally requires a balanced supply of up to 15 V, but I have arranged for it to be operated from a 9 V battery. The current drawn is about 2 mA and satisfactory operation is maintained with a supply down to 6 V with no noticeable degradation of the output signal.

The gain of the amplifier is defined by R_3 and R_4. R_3 is a thermistor which ensures that the output signal amplitude and frequency remain substantially constant at all operating frequencies.

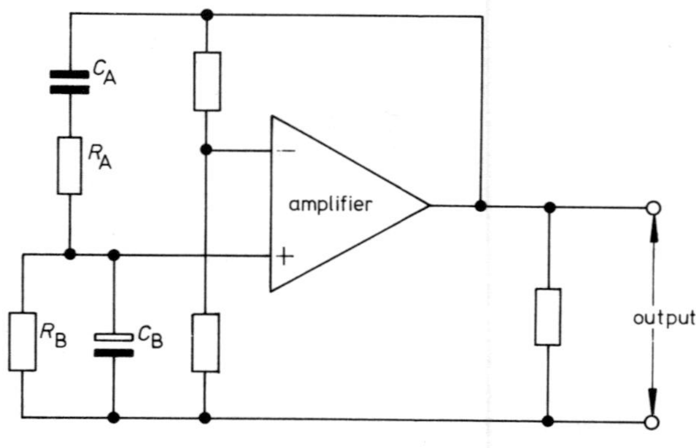

Figure 2.5

+9V
S2
output
0V
−9V
C_8 47μ
VR_2 5k lin
C_7 47μ
R_6 3.3k
R_5 3.3k
R_3 RA53
t°C
741
7
6
4
3
2
+
−
R_4 470
C_1 680n
C_2 68n
C_3 6.8n
×1
×10
×100
R_1 1k
VR_{1A} 10k lin
VR_{1B} 10k lin
R_2 1k
C_4 680n
C_5 68n
C_6 6.8n

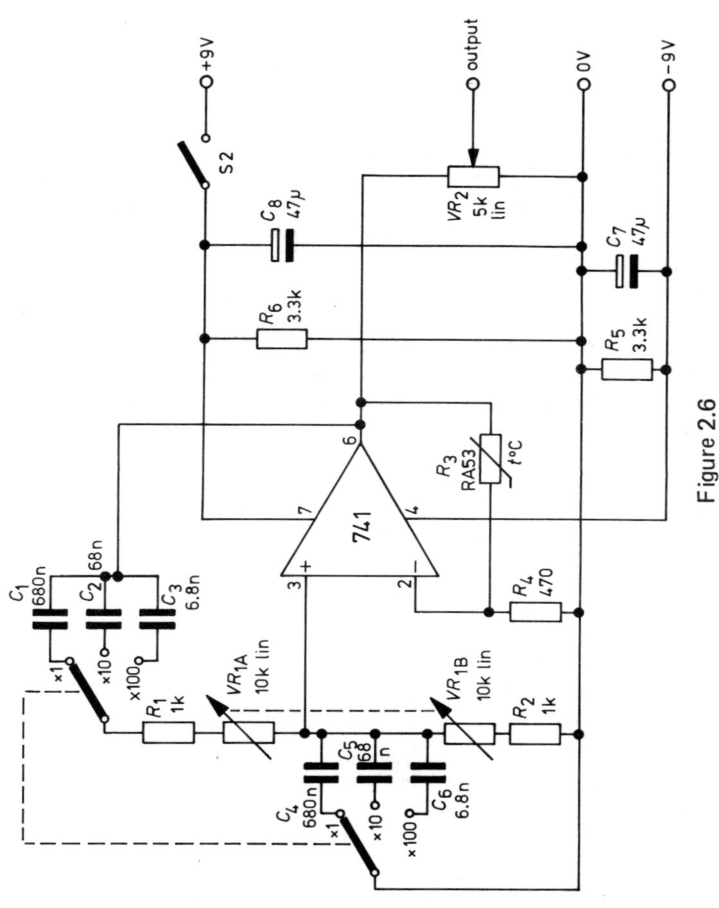

Figure 2.6

The frequency ranges are selected by a two-pole three-way switch. Capacitors C_1 and C_4 are for the lowest range (x1). C_2 and C_5 for the middle range (x10) with C_3 and C_6 for the highest range (x100). The frequency is continuously variable over each of these ranges by VR_1. VR_2 allows adjustment of the output voltage from 0 V to 4 V peak to peak.

Construction

A suitable Veroboard layout is shown in figure 2.7. The 741C that I used was in an 8-pin DIL package and I used a socket to mount it on to the circuit board. Although this adds to the initial cost slightly it may well avoid damage to the IC, which may occur if you try to solder it directly to the board. Don't fit the IC until construction is complete. I used a plastics box, 160 mm x 95 mm x 50 mm deep, (in which there is plenty of room for the circuit and the battery) and mounted S1, VR_1, VR_2, S2 with the output socket on the aluminium front panel. I soldered capacitors C_1 to C_6 directly between S1 and the circuit board providing support for one end. The earth side of output socket I connected via a solder tag and a piece of tinned copper wire to the common rail on the circuit board, thus providing support for the other end.

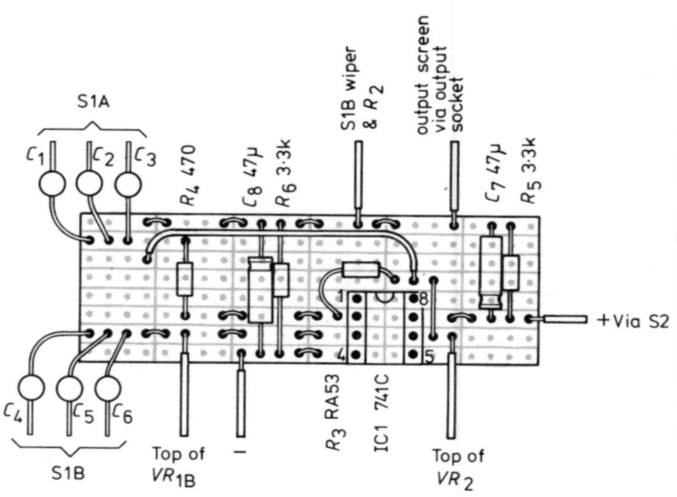

Figure 2.7

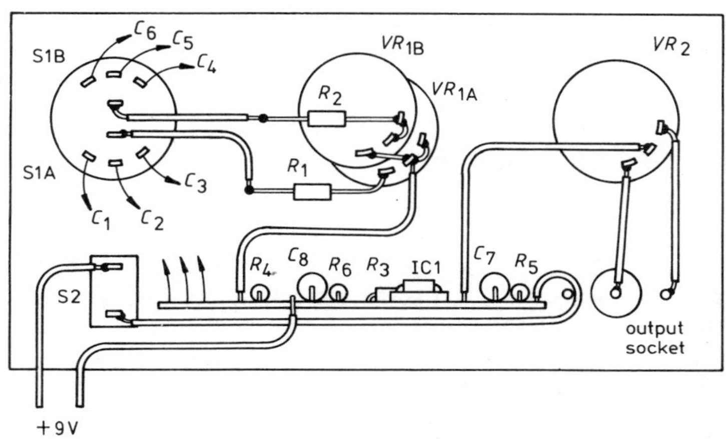

Figure 2.8 A.F. signal generator, rear view of front panel

If you look at my suggested front panel layout in figure 2.8, you will see that I mounted R_1 and R_2 between the relevant terminals of VR_1 and S1. When you have completed and checked all the wiring, fit the IC into its socket. *Make sure it is the right way round* (see figure 2.6).

Testing and Fault-finding

Connect the output socket to the input of an oscilloscope with its sensitivity set to about 1 V per division. Connect a PP3 or similar battery to the generator and switch it on. Switch S1 to the x1 position and turn VR_2 fully clockwise. Adjustment of your oscilloscope timebase should provide a good sinewave display. Turning VR_1 should vary the frequency of the output from about 20 Hz to about 230 Hz. Switching the range switch to the x10 position, VR_1 should now sweep between about 200 Hz and about 2.3 kHz. The top range with S1 in the x100 position should yield outputs of beween 2.0 kHz and 20 kHz. Over all three ranges the output voltage should be a constant 4 V peak to peak (approximately). If the generator is operating as it should, VR_1 can be calibrated. I suggest you mark positions on the front panel relating to the x1 range. I marked 25, 50, 100, 150, 200 and 225 Hz. If fairly close tolerance components are used for capacitors C_1 to C_6, these markings should be fairly accurate for the other two ranges. I then marked S1 positions x1, x10 and

100. Although I used a separate miniature toggle switch for S2, this could be replaced by a switch fitted to VR_2.

A Crystal Reference Oscillator

If you need to calibrate ac voltmeters, frequency meters or radio receivers, a very accurate reliable frequency standard is invaluable. Although this circuit does not approach the standards of the best commercial equipment, it is cheap and easy to build and provides a really stable, reliable source.

Circuit

The circuit diagram is shown in figure 2.9. I have used a 7400 TTL IC as the basis of the crystal oscillator. There are quite a few examples of this type of circuit around but this one, unlike some of the others, seems quite reliable. It uses two of the 7400 gates – connected as inverters – as the oscillator and a third gate as an output buffer. X1 is a plug in 1 MHz quartz crystal which, in this circuit, is connected in series mode. C_1 is included to adjust the fundamental oscillator frequency to exactly 1MHz. IC2 to IC7 inclusive are 7490 decade counters. Each of these divide the signal by 5 and 10, providing outputs of 1 MHz, 500 kHz, 100 kHz and so on down to 1 Hz. The outputs are nominal 5 V TTL squarewaves and because of the good risetime of these, they are rich in harmonics.

Construction

My Veroboard layout is shown in figure 2.10. I used sockets for the crystal and all the ICs. Treat the crystal gently because it will easily break if dropped. Fit all the resistors, the capacitors and sockets to the board first. Next make all the necessary interconnections – I used point-to-point wiring with PVC-covered flex. Plug in the crystal and then the ICs, making sure that they are the right way round.

One factor that can lead to oscillator drift is change of temperature. This is why, in critical applications, crystal oscillators are run in 'ovens'. This means that the whole unit is kept at a substantially constant temperature, irrespective of ambient temperature. I thought that such elaborate precautions were not justified for this application, but I did take one or two measures with a view to temperature stability. Firstly, I decided to use an external power supply because this can generate a certain amount of heat. Secondly, having fitted the circuit board into a suitable aluminium box, I filled the box with polystyrene packing pieces.

1Hz
5Hz
10Hz
50Hz
100Hz
500Hz
1kHz
5kHz
10kHz
50kHz
100kHz
500kHz
1MHz
IC7 7490
IC6 7490
IC5 7490
IC4 7490
IC3 7490
IC2 7490
IC1 7400
5V
C1 100n
R1 470
R2 470
CT1
X1

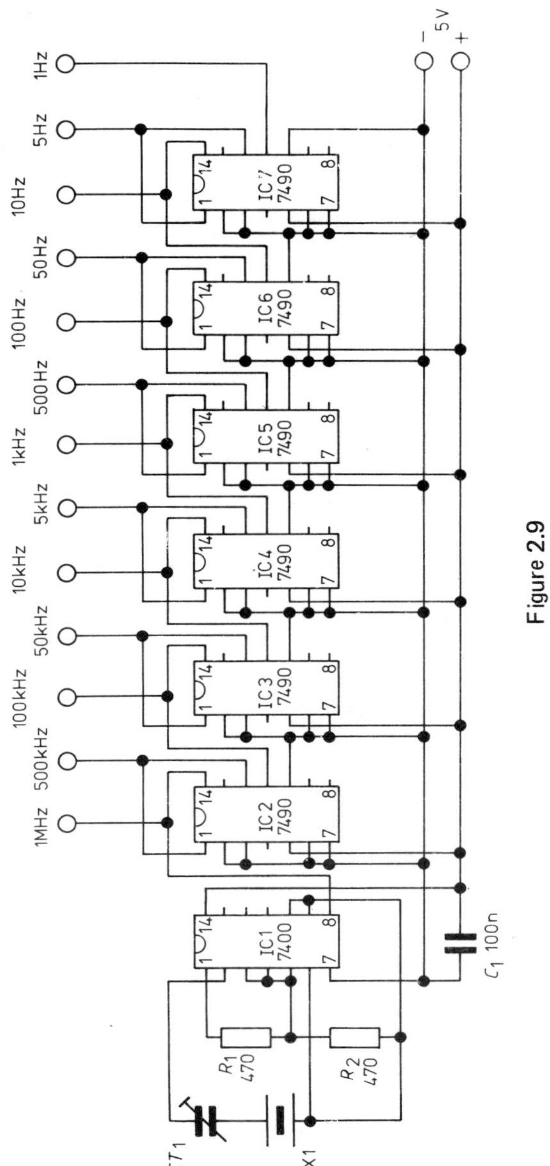

Figure 2.9

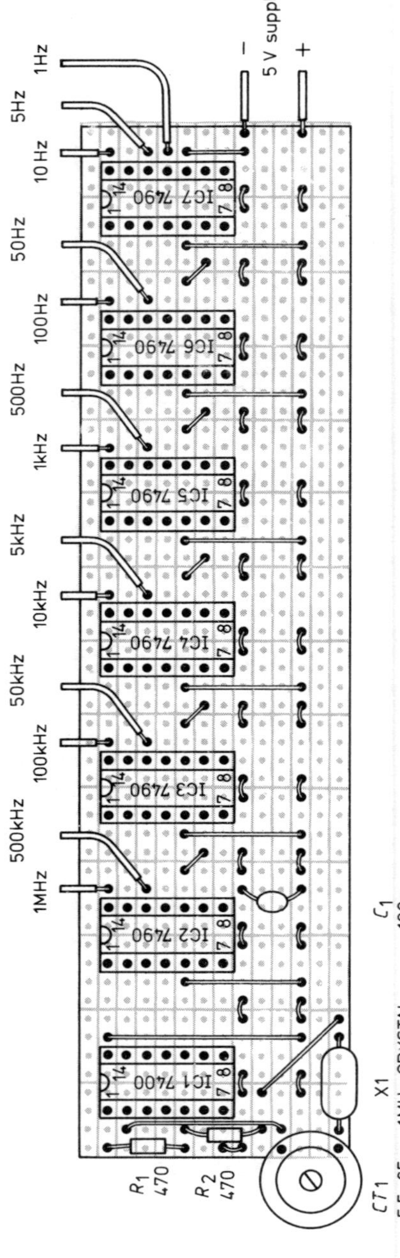

Note Interpin wiring is not shown in all cases. Check with the circuit diagram and interconnect the IC pins on the underside of the board using PVC-covered flex.

Figure 2.10 Crystal reference oscillator, circuit board layout.

Rather than use elaborate switching, with its associated unreliability, I used a separate 2 mm socket for each output. For the power supply input, I used a pair of 4 mm screw terminals – one red and one black.

Testing and Fault-finding

Having completed the construction of your crystal standard oscillator connect it to a stabilised 5 V power supply with an output current capability of about 200 mA.

Although crystals are extremely accurate, the oscillator will have to be calibrated, by adjustment of C_1, before it is ready for use. To do this, the easiest method is to check one of its outputs using a frequency counter or frequency meter of known accuracy. Another method is to compare one output of the oscillator with a known frequency standard, using Lissajous' figures. I have shown how to connect up for this in figure 2.11. Adust C_1 until the Lissajous' figure stops rotating. Access to accurately known frequency standards depend on where you live. In the United Kingdom, the BBC trans-

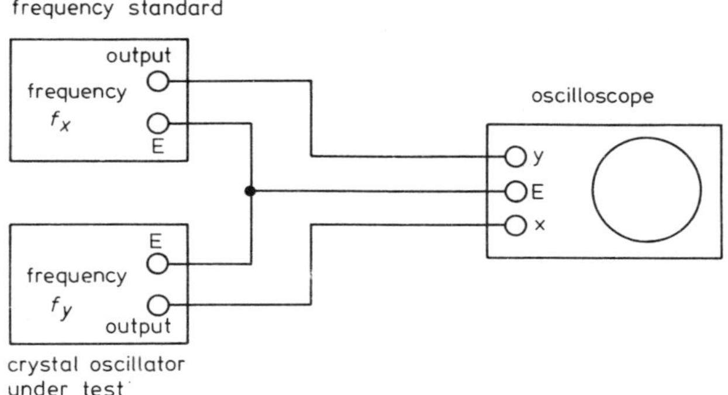

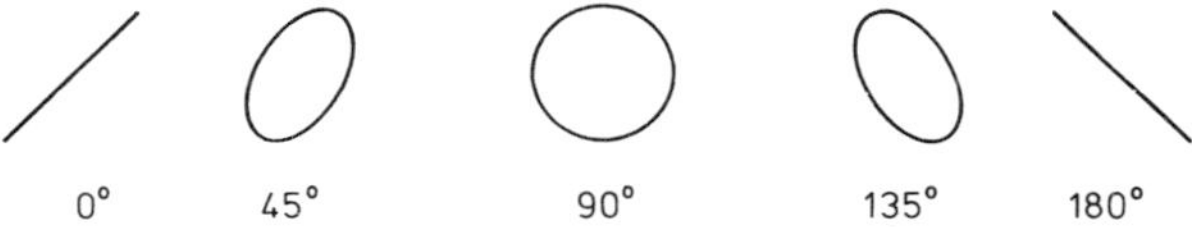

Figure 2.11 Test rig for Lissajous' figures. The actual shape of the display will depend on the phase relationship between f_x and f_y

mit I kHz and 400 Hz test signals during some parts of the day. Medium wave transmitters are normally within about ±0.05 Hz of their nominal carrier frequency, which should be good enough!

After calibrating the oscillator, check the frequency and waveform of each output using an oscilloscope. If there is a slight discrepancy between the theoretical output frequency and the measured one, remember that the crystal standard is probably more accurate than your oscilloscope — unless you have a very, very expensive one!

If you get no output at any frequency, switch off and check over your wiring again. No output at all indicates a fault in the basic oscillator. This circuit is a particularly reliable starter and so the most likely culprits are the 7400 and the crystal itself. Unplug the 7400 and try another one. If this does not work, I suggest you check the crystal by setting it up as I have shown in figure 2.12. Increase the generator frequency from just below 1 MHz and the low-impedance frequency of the crystal will be indicated by a sudden rise in the *Y* amplitude of the oscilloscope display. Keep increasing the frequency and you will get a sudden notch in the *Y* amplitude as the crystal reaches its high-impedance parallel frequency. If this proves correct, and assuming that you haven't misread the colour code on R_1 and R_2, then the trimmer capacitor is indicated as being faulty.

If you fail to get an output at a particular decade of frequency, then this indicates a faulty 7490. Before trying to get another one, confirm this by swapping the apparently faulty one for one that seems to be working normally. If the fault 'moves' with the suspected 7490 then it must be replaced but if the fault stays where it was, then you must have an undetected wiring fault.

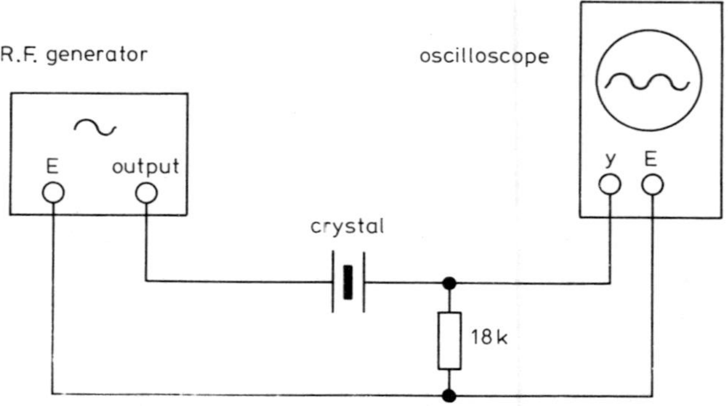

Figure 2.12 Test rig for checking crystals

A Waveform Generator

For many design and test procedures, a good-quality waveform generator is virtually indispensable. This one is based on the 8038 waveform generator. It produces sine, square or triangular outputs at any frequency between 20 Hz and 20 kHz. With the specified supply voltage, the output signal can be varied between 0 and about 5 V. A 741 op-amp is used as an output buffer which has the added advantage of output short circuit protection.

Circuit

In the circuit, shown in figure 2.13, IC1 is an 8038 waveform generator and IC2 is a 741 op amp — used here as a buffer. VR_2 is the frequency control and VR_1 is pre-set to give a minimum output frequency of 20 Hz. VR_3 sets the duty cycle of all three waveforms. This is adjusted to give an equal mark-to-space ratio of the squarewave output which corresponds to a duty cycle of 50 per cent. VR_4 and VR_5 are adjusted for best sinewave purity. It is possible to reduce sinewave distortion to almost 0.5 per cent by careful adjustment. Normally the sine, triangular and squarewave outputs are at different levels but R_6, R_7 and R_8 are included to equalise these. VR_6 is the output level control.

Pin 8 of IC1 is susceptible to pick up and C_1 is fitted to decouple it. C_2 is the timing capacitor that determines the frequency range over which the generator is tuned. For optimum frequency stability this should be the low temperature coefficient, silvered mica type specified. To achieve the 1000:1 variation of frequency, the timing resistors VR_1, VR_2 and R_1 must be returned to a supply that is a few mV higher than that connected to the duty cycle control (VR_3). This voltage difference is achieved by employing the forward voltage drop across D1. This is a small silicon diode such as an IN914 or IN4148.

C_3 and C_4 decouple the supply rails from high-frequency components. The supply itself may be derived from a pair of PP9 batteries or a mains power pack with up to ±15 V supply rails. S1 is a single-pole three-position rotary switch and is used to select the required waveform. S2 is a double-pole changeover switch used to switch the supply rails.

Construction

I have shown my Veroboard layout in figure 2.14. Use a 14-pin DIL socket for the 741. VR_1, VR_4, VR_5 and VR_7 are pre-set types and these must be mounted on the board.

C1 100 n
VR1 1k lin
VR2 10k lin
R1 18k
VR4 100k lin
VR5 100k lin
D1 1N914
VR3 1k lin
R3 4.7k
R4 4.7k
IC1
8038
R2 10M
C2 4.7n
R5 15k
R6 47k
R7 15k
R8 10k
S1
VR6 100k lin
IC2
741
R9 10k
VR7 10k lin
S2A
+9V
output socket
0V
S2B
−9V

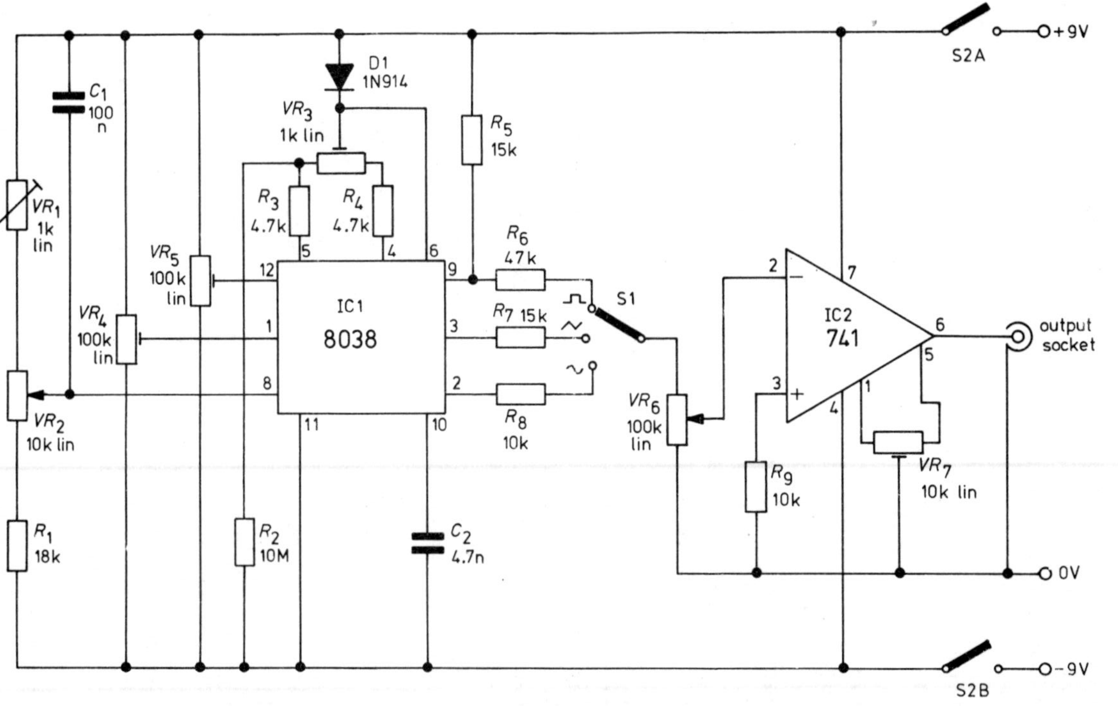

Figure 2.13

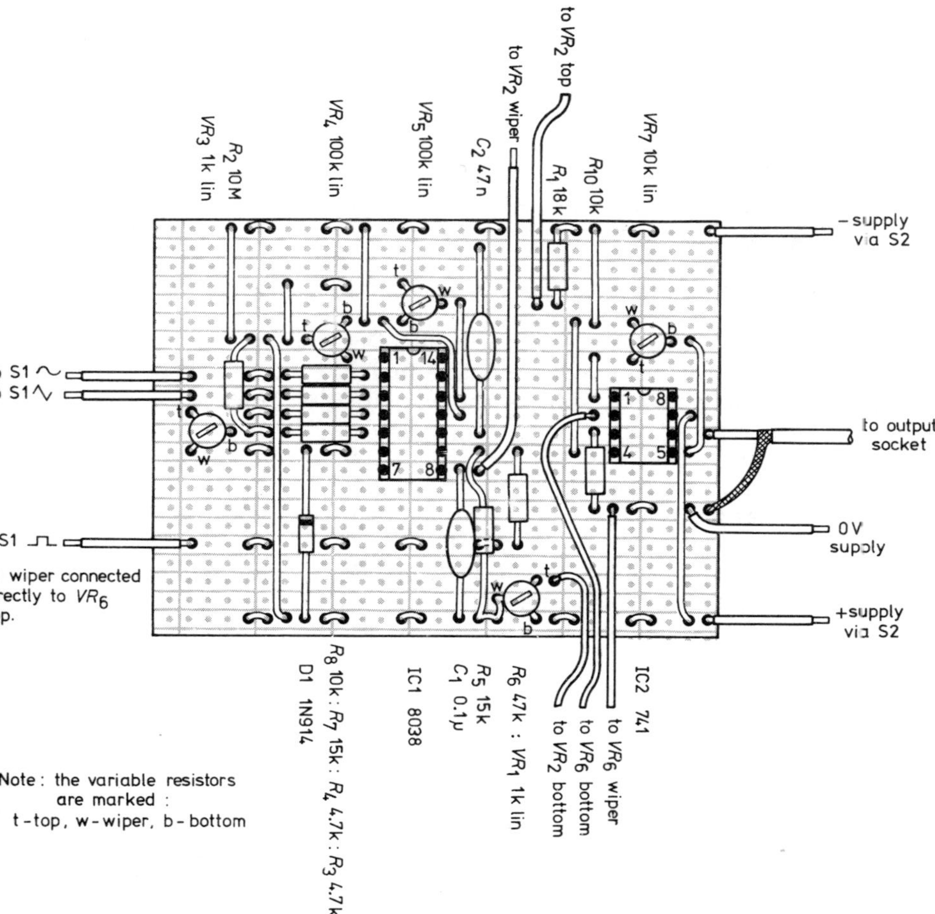

Figure 2.14 Waveform generator, circuit board layout. The variable resistors are marked T for top, W for wiper and B for bottom

The type of container that you use is largely dependent on your chosen power supply. If you choose PP9 batteries, then these must be accommodated. If you use an external, mains-operated power pack then a small container about 150 mm x 75 mm is quite adequate. VR_2, VR_6, S1 and S2 together with a suitable output socket – I used a co-axial type – must be mounted on the front panel.

Calibration and Testing

After constructing your generator and carefully checking the wiring (making certain the the ICs are the right way round) connect a power supply and switch on.

Connect an oscilloscope to the output socket and switch S1 to squarewave output. Adjust VR_3 until the squarewave displayed gives an equal mark-to-space ratio. Switch S1 to sinewave output and adjust both VR_4 and VR_5 for minimum sinewave distortion. With VR_2 set to minimum frequency adjust VR_1 for a frequency of 20 Hz. Variation of VR_2 should now change the output frequency from 20 Hz to about 20 kHz. Variation of VR_6 should change the output level from 0 to about 5 V – the maximum level is related to the supply voltage level. Disconnect the feed from S1 to the top of VR_6 and adjust VR_7 until the output from IC2 is as near to zero as possible. Replace the feed and the generator is ready for use.

Fault-finding

If the circuit fails to operate, try to isolate the faulty area. Disconnect the feed from the wiper of S1 and the top of VR_6. Using an oscilloscope, check pins 2, 3 and 9 of IC1 to see if you are getting any of the outputs. If not, then the 8038 is faulty. Before changing it, ensure that C_2 is not short circuit or D1 the wrong way round. If the signal outputs from IC1 are correct, then IC2 or its associated components must be faulty.

If you fail to get the full range of frequency variation – between 20 Hz and 20 kHz – then D1 may be short circuit.

A White Noise Generator

In the a.f. and r.f. equipment we spend most of our time trying to keep noise to a minimum. However, noise generators are very useful for detecting non-linearity in both a.f. and r.f. systems. The generator suggested here produces random or 'white' noise and is suitable for use with audio systems. It may also be used as a special effects generator, producing a wind-like noise.

Circuit

A number of electronic components produce noise, especially Zener diodes. The noisiest of all are those in the 12 to 40 V range with very small reverse currents flowing. In my circuit, shown in figure 2.15, I have used the emitter–base junction of a small silicon transistor (TR1) as a 'Zener diode'. The noise signal produced is ampli-

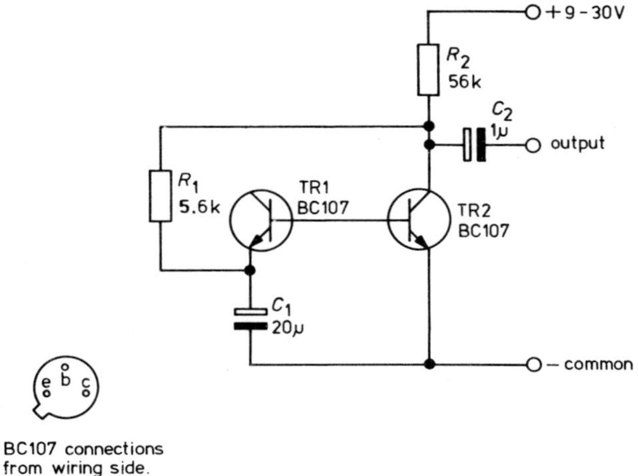

Figure 2.15

fied by TR2 and connected to the test circuit via the small coupling capacitor, C_2.

Construction

I constructed my generator on a small piece of Veroboard using the layout shown in figure 2.16. An ideal container is a cigar tube. There is no need to fit a battery as the circuit will tolerate a supply voltage

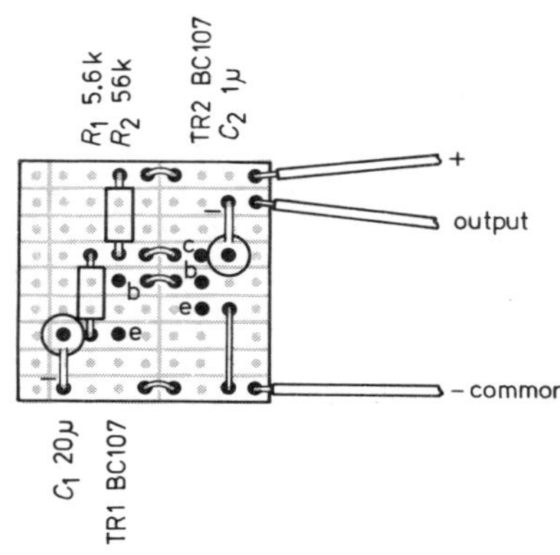

Figure 2.16

of anything from about 9 to 30 V. I fitted a pair of leads with croc-clips so that I could pick up a supply from any circuit under test.

Testing and Fault-finding

Connect the generator to a suitable supply with its output connected to an oscilloscope. You should be able to see a signal of several volts peak to peak (depending on the supply voltage) with a very broad spectrum of frequencies.

If you get no output, make the usual checks of the wiring — have you got the transistors the right way around? Is there any solder shorting across the tracks of the Veroboard? If all these checks find no errors, change each component in the order transistors, capacitors, resistors — checking the circuit after each change.

Stereo Balance

For maximum stereo effect the acoustic output from each loudspeaker should be balanced. Although your system may be electrically balanced, the siting of loudspeakers to fit into a ready furnished room may lead to an acoustic imbalance. This circuit provides an audio bleeping signal which may be used to achieve a perfect acoustic balance without the unpleasant ringing sensation produced by the continuous tone of a signal generator. The same circuit can be used as the basis of an alarm or special effects oscillator if you don't do very much audio work.

Circuit

The complete circuit diagram is shown in figure 2.17. It comprises a low-frequency multivibrator (TR3 and TR4), which is used to pulse the supply to a higher-frequency multivibrator (TR1 and TR2). The result is an output of about 1 s bursts of 500 Hz squarewave. The level of the output signal is controlled by VR_1 and the LED, D1, gives an indication that the circuit is operating — this will flash on and off in time with the slower multivibrator. The slower multivibrator should operate at about 0.5 Hz. Component types and layout are not critical but the value of R_8 was chosen for use with the type of LED specified. If you use a different one, the value of R_8 may need to be adjusted slightly. Although C_4 and C_5 are part of the timing circuit for the slower multivibrator, you can use the next preferred value either way since no electrolytic capacitor is manufactured to close tolerance. All transistors are general-purpose silicon *npn* types.

+
−
output
R_1 15k
R_2 330k
R_3 15k
R_4
C_1 10n
C_2 10n
C_3 10n
VR_1 10k
R_5 390
C_4 330μ
R_6 2.2k
R_7 2.2k
C_5 330μ
R_8 390
D1 TIL209
TR1
TR2
TR3
TR4
R_9 56
wirewound
e b c
transistor from wiring side

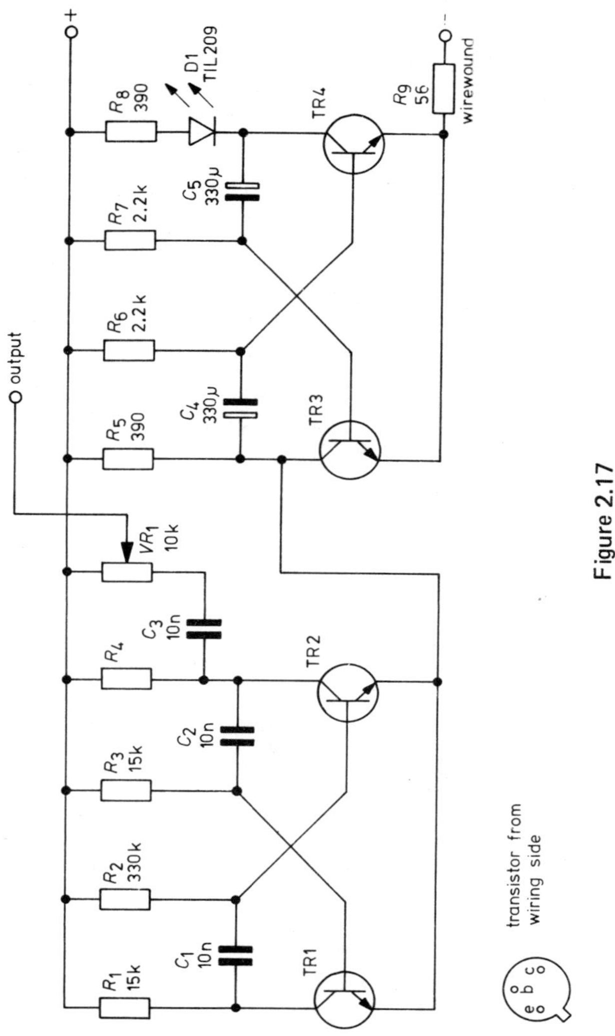

Figure 2.17

Construction

I fitted all the components to a single piece of Veroboard (see figure 2.18) standing them up to conserve space. I didn't think it was worth using a box for this unit because it is only going to be used very occasionally. I took the output, via a length of single-screened lead, to a DIN plug wired to common the stereo inputs, which suits my amplifier. Perhaps it would be a good idea to include a second out-

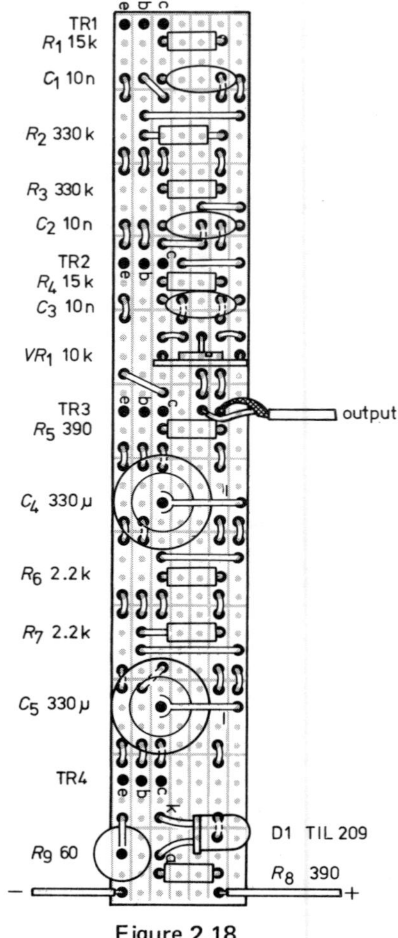

Figure 2.18

put lead fitted with phono plugs since many Japanese amplifiers use these.

I used a pre-set skeleton pot for VR_1 so that I could mount it directly on to the circuit board. A small, printed board mounting (spindle type) would be equally suitable. The circuit can be run from a single PP3 battery so I fitted a suitable battery clip for this. You can fit an on/off switch if you like but, since the LED is flashing when the unit is working, it is virtually impossible to leave it switched on by mistake.

Testing and Fault-finding

Plug the unit into your amplifier input, choosing one of the less sensitive sockets (such as tuner or tape). Connect a good battery or 9 V dc power supply and switch on the amplifier. The LED should start start flashing and you should hear a steady bleeping tone coming from your loudspeakers. You should adjust VR1 to give a suitable level of output. Now adjust the amplifier stereo balance control until the bleeps seem to be coming from a point midway between the loudspeakers.

If the unit fails to operate, assuming that the battery is working first try to isolate the faulty area. For example, if the LED is flashing but no bleeps are heard, then the fault must lie either in the higher-frequency multivibrator or in the output connections. If you have an oscilloscope available, a quick check of the collectors will reveal whether the oscillators are operating. Check your construction carefully, looking for unsoldered connections or short circuits between the tracks on the board. If you have used a DIN plug, have you connected the screened lead to the correct pins? These are usually 2 (screen) 3 and 5 (signal) but they may be different for your amplifier. If you become convinced that the fault must be a component failure, change each component starting with semiconductors, then capacitors and lastly resistors, testing the unit after each alteration.

3

Test Meters

A dc Multimeter

There are many reasonably priced, commercial multimeters available and you would be justified in wondering whether it is worth building one. At the heart of many multimeters is a good-quality sensitive moving coil movement and one of these can cost the constructor as much as a ready built measuring instrument. However, if you already have a movement or you can shop around for a good quality 'surplus' one, then I think it is worth having a go at constructing your own dc multimeter.

My circuit will indicate dc volts and current and has a range to measure resistance. It uses a 0 to 100 μA moving coil movement, with an internal resistance of 1300 Ω and a scale marked both 0 to 100 and 0 to 30. However, I think I have included enough detail to enable you to accommodate virtually any good-quality movement. All the resistors that I have used are standard values from the E24 series (although I recommend that you use ± 1 per cent tolerance, high-stability components where possible).

Circuit

The complete circuit is shown in figure 3.1. I have kept switching to a minimum, using only a single-mode switch (S1). The various

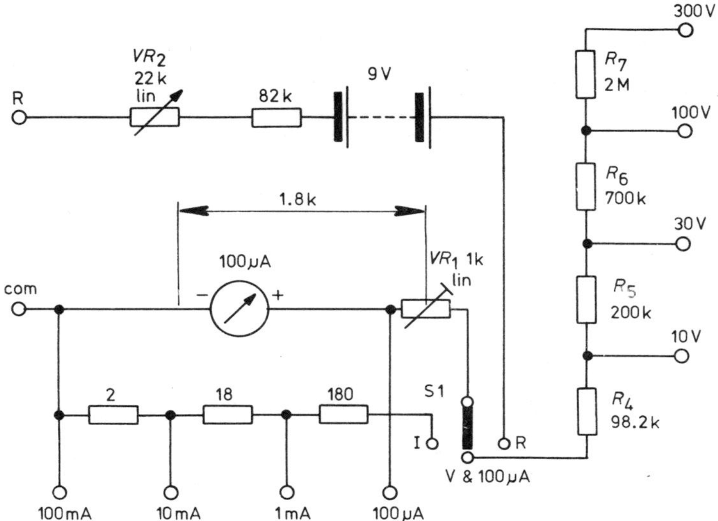

Figure 3.1 DC multimeter, complete circuit layout

ranges are selected by plugging the test leads into the appropriate sockets (with one socket common to all ranges). VR_1 is fitted so that the effective internal resistance of any movement fitted can be adjusted to a standard 1.8 kΩ. My meter movement is rated at 1300 Ω ±15 per cent. This degree of tolerance is typical – the best you could expect would be about ±10 per cent – but I have never come across a 100 μA movement with a greater internal resistance than 1.8 kΩ and this is a convenient value to use as I will explain later.

The lowest current range (100 μA) is obtained by switching S1 to the '100 μA and V' position and connecting directly across the meter movement. This is the only function that excludes VR_1. For the remainder of the current ranges, S1 is switched to the 'I' position. R_1, R_2 and R_3 form a 'universal shunt' and their values are calculated as follows.

First calculate the shunt needed for the next range up from the meter movements full-scale deflection (1 mA in my circuit). This is 200 Ω – for the calculations and a schematic diagram, see figure 3.2. When a test lead is plugged into the 1 mA socket and 1 mA is applied, 100 μA flows through the meter movement and the remaining 900 μA flows through the 200 Ω shunt. Notice that 1.8 kΩ

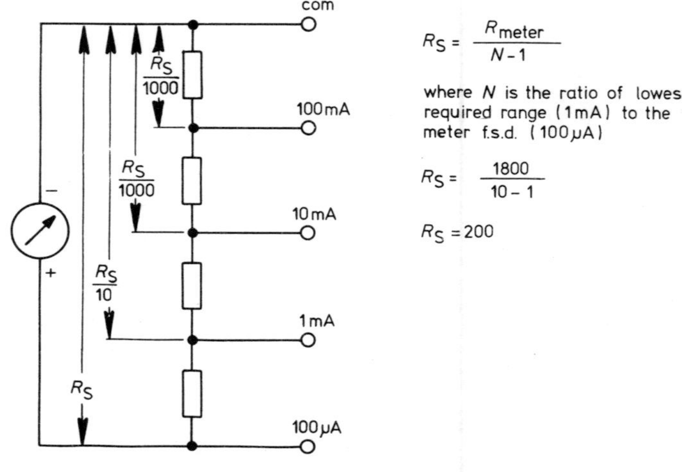

Figure 3.2 Calculation of universal shunt

is exactly 9 x 200 Ω and the 200 Ω is the sum of R_1, R_2 and R_3 all of which are standard value resistors.

When the test lead is moved to the 10 mA socket and 10 mA is applied, the shunt is now 20 Ω ($R_2 + R_3$) and the meter resistance is now 1980 Ω (1.8 kΩ + R_1). This means that 100 μA will flow through the meter and the remaining 9.9 mA flows through the shunt.

When the 100 mA range is selected and 100 mA applied, the shunt is 2 Ω and the meter resistance 1998 Ω (1.8 kΩ + R_1 + R_2) resulting in 100 μA flowing through the movement and 99.9 mA flowing through R_3.

R_1 and R_2 should be ±1 per cent tolerance, high-stability components. This type is not generally available with a value of 2 Ω (R_3), and so I suggest you use a small wire-wound component because these are usually made to close tolerances anyway.

The voltage ranges are used with S1 in the '100 μA and V' position. If the 10 V range is selected and 10 V applied, a series resistor must be added to the effective meter resistance (1.8 kΩ) in order to limit the current to 100 μA. The total resistance that must be presented is 100 000 Ω (from $R = V/I$). Hence R_4 will be 100 kΩ − 1.8 kΩ = 98.2 kΩ. This is constructed from three standard resistors in series: 43 kΩ + 47 kΩ + 8.2 kΩ.

If the 30 V range is selected and 30 V applied, a total resistance of 300 kΩ must be presented by the meter. R_4 and the meter movement total 100 kΩ, so R_5 must be 200 kΩ.

The 100 V range requires a total resistance of 1 MΩ and the 300 V range needs a total of 3 MΩ. Hence R_6 is 700 kΩ (430 kΩ and 270 kΩ in series) and R_7 is 2 MΩ.

The same theory can be used to extend the ranges or incorporate intermediate ones as you require. All resistors (R_4 to R_7 inclusive) should be ±1 per cent tolerance, high-stablility components.

To measure resistance, S1 is switched to the 'R' position. The circuit simplifies down to a 9 V battery and a 90 kΩ resistor in series with a meter movement assumed to have zero internal resistance (see figure 3.3). If the R and common terminals are shorted together, this will result in a full-scale deflection reading being obtained. Any value of resistance included between the terminals will result in a proportional deflection. VR_2 is used to adjust for zero over a wide range of battery conditons – a new PP3 will give about 9.5 V but adjustment of VR_2 allows the battery to fall to about 8.5 V before it has to be renewed. The resistance scale will give a direct reading of resistance from about 1 kΩ to about 5 MΩ.

Construction

I mounted my meter movement, together with S1, VR_1 and the input sockets, on to a metal front panel about 350 mm x 75 mm which I fitted to a 50 mm deep box. A piece of Veroboard makes a suitable mounting for the shunt and multiplier resistors. Although modern metal film resistors are less sensitive to heat than older types, use a croc-clip as a heatsink when you solder them – this will avoid any possibility of their changing value through being heated.

I used screw terminals that will accept a 4 mm plug for my range sockets. They are available in a variety of colours – I used a black one for the common terminal, red for the voltage terminals, yellow

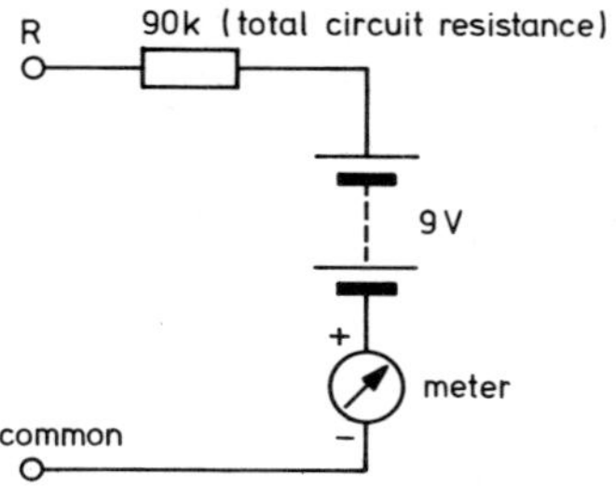

Figure 3.3 Simplified circuit on resistance range

for the current ranges and a green one for resistance. It makes a really professional job if you can use a similar colour code for the switch and meter panel signwriting.

Don't worry if you can't get a meter movement with the kind of dial marking you require, the problem is usually quite easily solved. Most dial panels are made of aluminium or plastics material and are only held on by a couple of small screws. Undo these and you can remove the panel. You will find that the back is blank and ready for your markings. A plastics panel will be white already and can be marked with a stencil or using Letraset dryprint. Aluminium panels usually have a dull silver etched finish – you can use this or spray

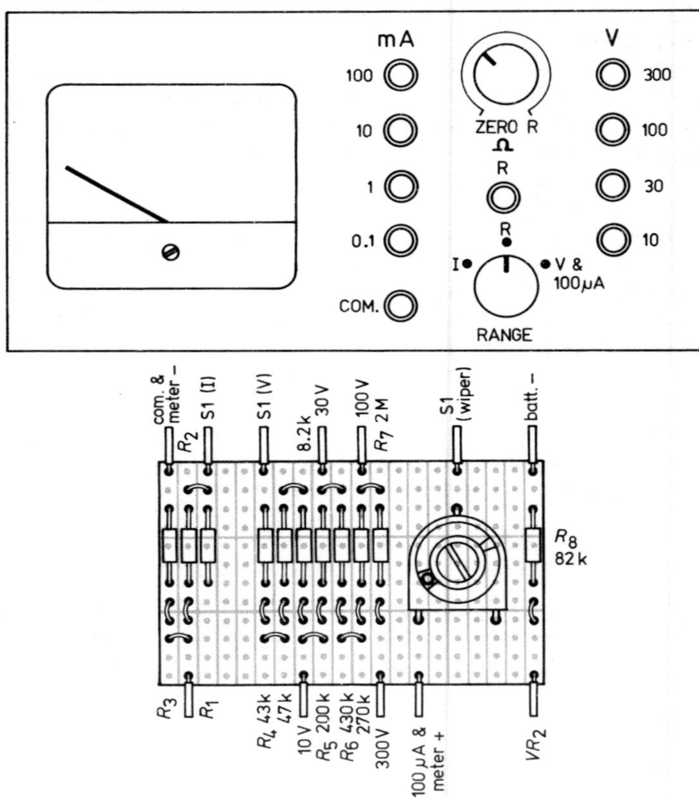

Figure 3.4 DC multimeter, front panel and printed board layouts

it any colour you like, marking it in the same way. I do not recommend that you spray a plastics panel as some paints attack some plastics giving you a very individual finish!

I have shown my own front panel and Veroboard layouts in figure 3.4, but these are only a general guide which you must adapt to the components that you use.

Calibration and Testing

After you have assembled the multimeter and checked all the wiring, it must be calibrated.

Firstly you must set up VR_1, and to do this you will need an ordinary carbon track linear variable resistor (VR_3) of about 25 kΩ and another meter (B) capable of registering 1 mA f.s.d. You will need a 9 V supply as well, but you can use the PP3 battery from the meter for this.

With VR_3 set to maximum initially and the multimeter set to read 1 mA, connect all the components in the test circuit shown in figure 3.5. Adjust VR_3 to give a reading of 1 mA on meter B. Now adjust VR_1 to give 1 mA (f.s.d.) on the multimeter. Adjust VR_3 and VR_1 in turn until both meters read exactly f.s.d. (1 mA). The multimeter can now be removed from the test circuit and the battery replaced. VR_1 is now set up, and this automatically calibrates the voltage and current ranges.

One way that you can calibrate the resistance range is to measure a set of accurately known resistors — a decade box is ideal for this — and mark the meter scale accordingly. I think that this would be a

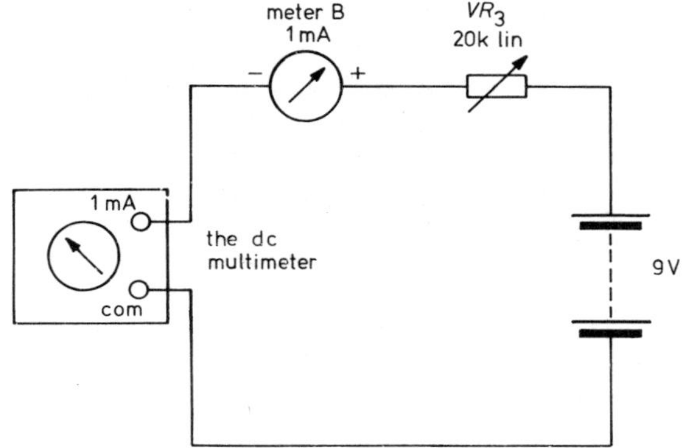

Figure 3.5 The test rig to set up VR_2

bit tedious, especially since you must frequently check the battery condition by zeroing the meter. I think a better way is to calculate points on the current scale (0 to 100 μA) corresponding to particular values of resistance, using the equivalent circuit shown in figure 3.3 Any resistor placed across the test terminals is added to the 90 kΩ series equivalent resistance and then the current is calculated using the formula $I = V/R$. For example: a test resistance of 10 kΩ corresponds to

$$I = \frac{V}{R}$$

$$I = \frac{9}{90\ \text{k}\Omega + 10\ \text{k}\Omega}$$

$$I = \frac{9}{100\ \text{k}\Omega}$$

$$I = 90\ \mu\text{A}$$

A few more calculations will reveal that 90 kΩ corresponds to 50 μA, 800 kΩ corresponds to just over 10 μA, and so on. You can soon build up to a comprehensive calibration chart and then mark your scale accordingly. This method avoids any error introduced by a failing battery, but if your calculating skills are as good as mine, I suggest you double check them – twice!

Fault-finding

If the circuit fails to function in any respect and assuming that your meter movement is a sound one, check your wiring again. If you are getting wrong indications on voltage or current ranges, then you may have made a mistake when reading the colour codes of the resistors.

Failure of the meter to zero on resistance range means that the battery is flagging – you need a fresh one.

If, when measuring known voltage around a circuit, the meter reads low, then this is not actually a 'fault' but more a shortcoming of almost all practical voltmeters. Don't forget that your multimeter takes a certain amount of current from the circuit that is being measured – as much as 100 μA – and this is customarily expressed as a resistance per volt, in this case 10000 Ω/V.

In figure 3.6, the voltage across R_B should be about 5 V but your multimeter will indicate only about 3.3 V when you use the 10 V range. The meter's resistance on this range is 100 kΩ. This is in parallel with R_B, reducing its effective value to 50 kΩ. The available

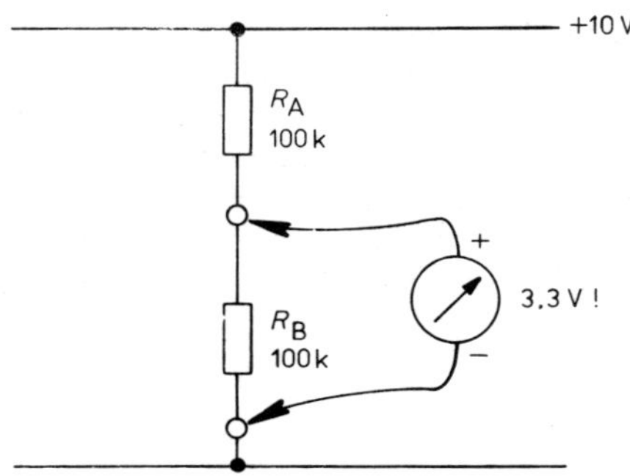

Figure 3.6 The effect of a voltmeter on a circuit

10 V distributes itself through the circuit in proportion to the resistance of each circuit element, thus about 6.6 V is dropped across the 100 kΩ of R_A and the remaining 3.3 V is dropped across the 50 kΩ of R_B and the meter in parallel.

To minimise this effect, you should select a range where the resistance of the multimeter is at least 10 x the resistance of the circuit being measured.

High Input Impedance Voltmeter

The trouble with most multimeters is that, when used as voltmeters, they load the circuit being measured. A 20000 Ω/V meter draws 50 μA and, on the 1 V range, it has an input resistance of only 20 kΩ. Even on fairly low impedance transistor circuitry, such a meter will read low, no matter how accurate it is.

This circuit uses a single BIFET op-amp as a meter amplifier resulting in an input resistance of more than 10 MΩ. With a resistance as high as this (comparable with many commercial electronic voltmeters) the meter will have very little influence on the circuit being measured — it draws only 0.1 μA on the 1 V range!

Circuit

The circuit diagram is shown in figure 3.7. Resistors R_1, R_2, R_3 and R_4 are multipliers providing four ranges, switched by S1. You should use ±1 per cent tolerance, high-stability resistors for R_1 to R_4 inclusive. For S1, I used a two-pole four-position rotary switch, connecting the two poles in parallel for extra reliability. Diodes D1 and D2 protect the op-amp from overload damage. VR_1 is used to null the

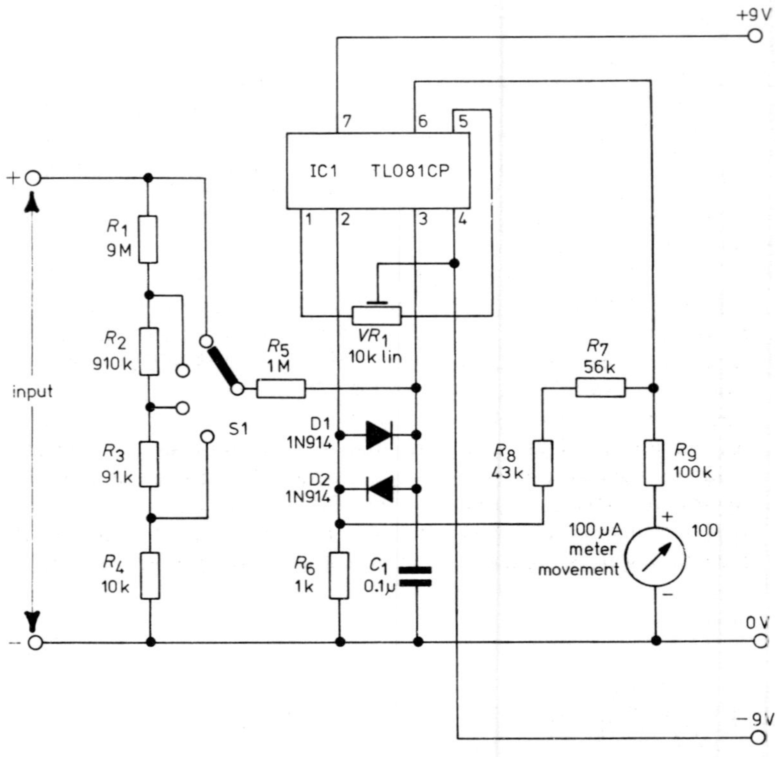

Figure 3.7

input offset voltage of the op-amp and provides a means of zeroing the meter for no input. C_1 ensures that the amplifier remains stable under all conditions and its gain is defined by R_6, R_7 and R_8 (these components together with R_9 should be ±1 per cent tolerance, high-stability types).

The amplifier itself is a TLO 81CP which is pin-compatible with the 741C op-amp. However, it offers several improvements over the latter, the most important for this application being its much higher input impedance: 1 million MΩ!

The meter movement can be any reasonably good-quality moving coil movement with an f.s.d. of 100 μA.

Construction

I mounted the meter movement, S1 and a pair of screw terminals capable of taking a 4 mm plug on to a front panel about 150 mm x 75 mm. With a very high input impedance amplifier, pick up from hand capacitance can be a problem. I used a switch with a plastics spindle for S1 and left plenty of clearance around the terminals of the input sockets to minimise this effect. I used a pair of PP3 batteries to supply the amplifier and fitted a miniature double pole toggle on/off switch to my front panel.

A suitable Veroboard layout is shown in figure 3.8. I used an eight-pin DIL socket for IC1, but if you want to mount it directly on the board, short all of its pins together using a small bulldog clip while you are soldering. This acts as an extremely effective heatsink. You should not, in any case, fit the IC until last and you must be careful to fit it the right way round. Leave the leads of the diodes full length, forming them into small loops around a screwdriver blade for compactness (see figure 3.9). The PP3 batteries will fit inside the box provided you use one at least 50 mm deep.

Your meter scale should already be marked 0 to 100, and it should only be necessary to mark it with a V (using stencils or Letra-

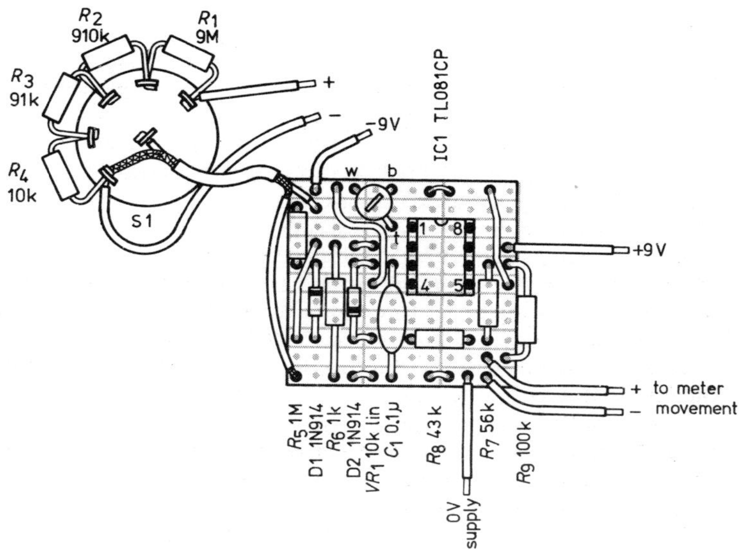

Figure 3.8

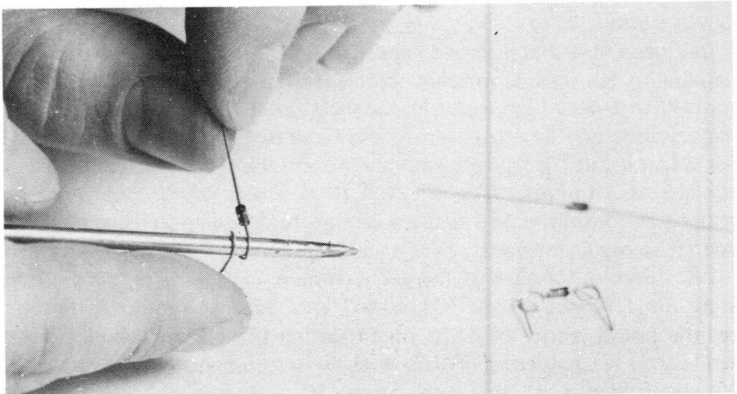

Figure 3.9 This illustration shows the way in which the diode lead is formed using a small screwdriver.

set dry print) unless, of course, you can get hold of a meter already so marked.

Setting-up and Testing

When you have constructed your voltmeter, and carefully checked the wiring, check again to see if the IC is the correct way round.

Before switching on, short the input terminals together. This avoids misleading effects of hand capacitance – part of C_1's job is to decouple any 'hum' signal picked up by the input, but it is just as well to be certain. When you switch on, you will probably get a deflection on the meter despite having no input. You should be able to zero the meter by gently adjusting VR_1

Remove the short from the input terminals – you may get a small pulse on the meter as you do this, but don't worry, it's quite normal.

The instrument requires no further calibration but should for peace of mind be checked against some known standard. You can, for example, measure the fixed output voltage of a stabilised power supply. You could compare the performance of your instrument with that of a good-quality commercial one – if you can borrow one. Connect up the meters as I have shown in figure 3.10. Set the power supply for several convenient readings on your voltmeter and, for each one, record the corresponding reading on the commercial meter. The main factors controlling the accuracy of your instrument are the selection of the resistors R_1 to R_9, although the tolerance of R_5 is not important and the accuracy of the meter movement itself.

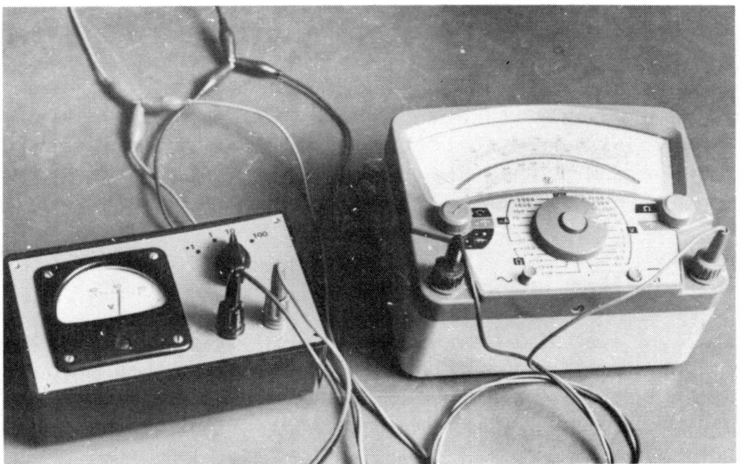

Figure 3.10 The completed test meter being compared with a good quality reference meter. Note the way the leads are connected in parallel, both meters reading the same voltage source.

r.f. Probe I

This simple probe may be used, with a dc electronic voltmeter, to determine the level of r.f. voltage in a circuit under test. The circuit is shown in figure 3.11. C_1 isolates the diode from any dc level present in the test circuit – use a 470 pF disc ceramic capacitor be-

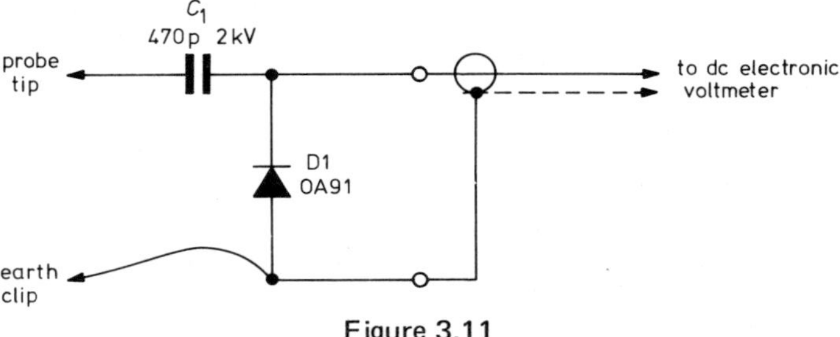

Figure 3.11

cause even a 2 kV working component is still very small. D1 is an OA91 which rectifies the r.f. signal. The indication on the dc scale of an electonic voltmeter will be $\sqrt{2}$ x r.m.s. value of the r.f.

Use a container such as a cigar tube, with the probe tip insulated from the metal casing. The output to the electronic voltmeter can be taken via a single-screened lead, using the screen to earth the metal container.

Measuring Meter Internal Resistance

Have you ever tried measuring the internal resistance of a moving coil movement using an ohmmeter? Were you annoyed when the moving coil movement's pointer bounced up against the end stop? Try using the circuit that I have shown in figure 3.12.

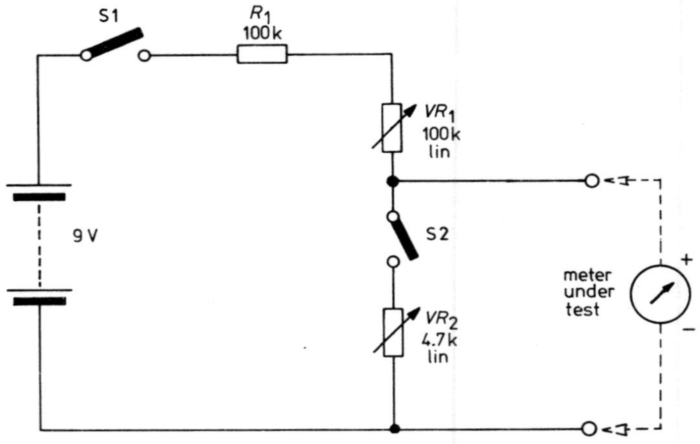

Figure 3.12 Test layout

Using the Circuit

Set VR_1 to maximum, open S2 and close S1. Adjust VR_1 until the meter under test indicates f.s.d. Now close S2 and adjust VR_2 until the meter indicates half full scale. The resistance between the slider of VR_2 and its bottom end is now equal to the internal resistance of the meter under test. Remove the meter, open S1 and measure this resistance with an ohmmeter.

The values I have given should be suitable for any meter up to a sensitivity of 50 μA. You could use a 9 V stabilised power supply in place of the battery, particularly when less sensitive movements are being tested.

4

Component Checkers

Estimating Unknown Capacitors

If you connect up the equipment as I have shown in figure 4.1, you will be able to make a pretty close estimate of the value of C_X, an unknown capacitor. The X and Y gains of the oscilloscope must be equal and the output from the generator must be fully floating.

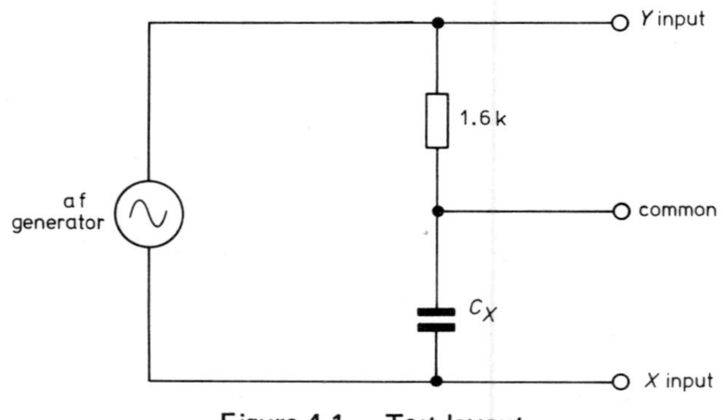

Figure 4.1 Test layout

Adjust the generator frequency until a stationary Lissajous' figure, as nearly circular as possible, is obtained. The value of C_X is given by

$$C_X = \frac{1}{f \times 10^4}$$

where f is the generator frequency. The technique is limited to reversible capacitors, of course.

A Simple Capacitance Bridge

Over a period of time, most constructors accumulate a large collection of components. Resistors have an almost universal colour code and their value can easily be confirmed using a multimeter. Capacitors are not so easy. These either have their value printed on them or they are marked with one of a bewildering variety of colour codes. The printed values are often worn off with handling and I, for one, have trouble remembering with certainty all the different colour codes. You can save a lot of money by buying capacitors in large bags marked 'out of specification', but you must check them before use. I think it is good policy to check all components before you use them, as this will save a lot of time later.

You can use a variety of methods to measure capacitance. An ordinary voltmeter or an oscilloscope can be used, but these methods involve calculations. The circuit that I suggest uses the bridge method and gives a direct reading of capacitors between a few pF and about 10 μF with adequate accuracy for general workshop use.

Circuit

The basic schematic diagram is shown in figure 4.2. The bridge is balanced by adjusting the ratio arms P and Q. At balance the detector will indicate a null point (a point of almost zero signal).

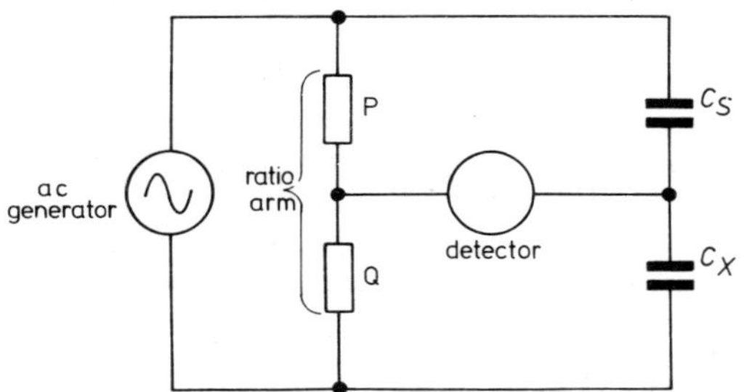

Figure 4.2 Simple capacitance bridge, schematic diagram

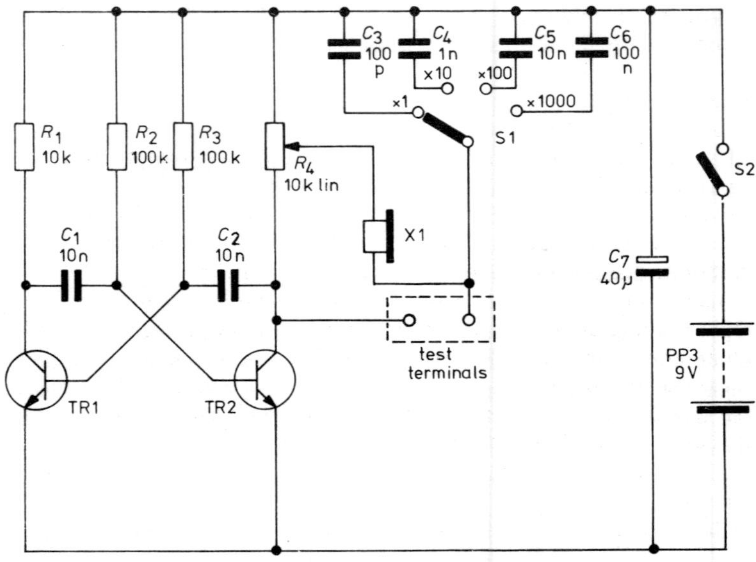

Figure 4.3

In my practical circuit shown in figure 4.3, the generator used is a simple low-frequency multivibrator since, unlike most direct-reading capacitance meter designs, waveform error is not a problem. The ratio arms P and Q are formed by a single variable resistor (TR2 collector load, R_4). C_s is one of four known capacitors of 100 pF, 1 nF, 10 nF and 100 nF. S1 selects the appropriate capacitor to give ranges of x 1, x 10, x 100 and x 1000 respectively. TR1 and TR2 may be almost any pair of a.f. transistors. The frequency of oscillation is set by C_1, C_2, R_2 and R_3. In my circuit this is about 650 Hz but any frequency between about 500 and 1500 Hz will work equally well with my suggested null detector – a crystal earpiece. The accuracy of the device is largely dependent on using close tolerance components for C_3, C_4, C_5 and C_6. The value of C_7 is not critical, however, because I chose it as a compromise between smoothing performance and reasonable compactness. For long life and reliability, the switch S1 should be a good-quality component. I used a two-pole four-position rotary switch connecting the two poles in parallel. R_4 must be a linear carbon-type potentiometer since these vary continuously. A wire-wound type varies in small steps and it may prove difficult to obtain a null point. The effects of hand capacitance will be minimised if you use a potentiometer with a plastics spindle. I used one with a built-in switch for the power supply.

Construction

I built the oscillator and range-selection capacitors on to a small piece of Veroboard using the layout shown in figure 4.4. You will need a box of about 150 mm x 75 mm to contain the bridge. I built mine into a plastics box with an aluminium front panel and figure 4.5 shows a suitable front panel layout for mounting R_4, S1 and the test terminals. I didn't think it was worth fitting a socket for the crystal earpiece because this would be another source of unreliability. I wired mine directly into the circuit bringing it out through a small Neoprene grommet in the front panel. For the test terminals I used a double spring-loaded trigger terminal – one red, one black. This has given reliable service for several months now. I powered my bridge with a PP3 battery mounted in the box.

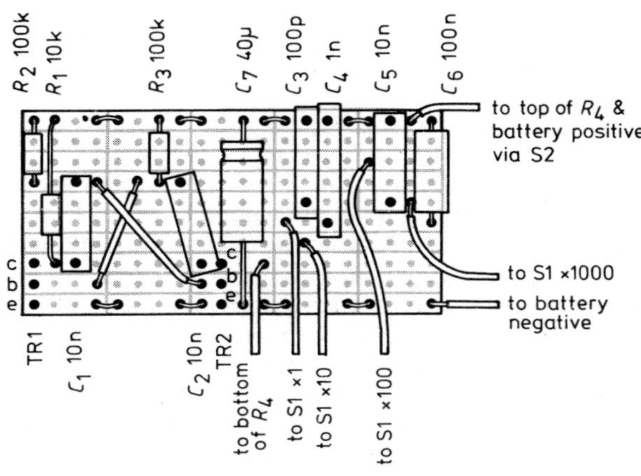

Figure 4.4

Testing and Calibration

Having constructed the circuit, carefully check your wiring and then connect a good battery and switch on. With no test capacitor connected, a buzz should be heard in the earpiece and you should be able to adjust its volume using R_4.

Obviously the bridge can be calibrated against a commercial bridge, but on the assumption that you don't have access to one, I suggest you use the following procedure.

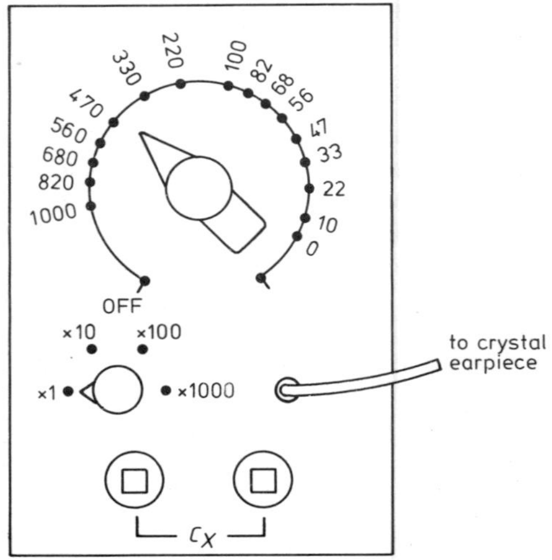

Figure 4.5 Simple capacitance bridge, front panel layout

First use a paper label to make the calibrated scale (this can be improved later, if you like, with a system like Letraset print.) Then mark the ends of the pointer travel and find the null point – the point of minimum signal, remember – near the top end of the dial. This is the zero point, so mark it.

Connect a close tolerance 100 pF capacitor (silvered mica types are generally suitable) across the test terminals. With the range switch (S1) in the x 1 position, find another null point and mark it 100. Change the range switch to x 10. This should give a null further up the scale. Mark this as 10. Next, substitute a close tolerance 1000 pF capacitor for the 100 pF one. With the range switch on x 10 this should give a null at the '100' point already marked. This gives a cross-reference for the previous test. With the range switch at x 1, a null should be obtained further down the scale. Mark this as 1000.

Continue with this technique using as many calibration capacitors as possible. If you can get hold of one, a precision capacitance substitution box is ideal for this procedure. If you are using separate components, you can make up intermediate values by constructing series and parallel combinations.

The null obtained at each step of the calibration procedure should be sharp and well defined. You must use a crystal earpiece because a moving coil type will not work very well – if you are in doubt as to which type you have, check it with an ohmmeter. The crystal type will produce a small click when the meter is connected but will appear open circuit. The moving coil will give a lower resistance reading.

If you try to measure small electrolytic capacitors, you will find that the inherent leaky nature of this type will give a poorly defined null. If you get this result with any other type of capacitor, this indicates that the component is excessively leaky and should be destroyed.

Fault-finding

If the circuit fails to operate, check that you are not using a flat battery. Next, give your wiring another check, particularly the connections to S1. If this is all right, check whether the oscillator is working. To do this, short the test terminals together with R_4 in about the centre of its travel. A buzz should be heard from the earpiece. You can confirm that the earpiece is working by connecting it across a 1.5 V battery when it should click loudly. If the oscillator is not working, replace the transistors, capacitors and resistors in that order, checking the circuit after each change.

Electrolytic Capacitor Tester

The use of bipolar transistors in modern electronic equipment has resulted in low-voltage high-value electrolytic capacitors being commonplace. These devices are manufactured to very wide tolerances and they need to be reformed if they have been stored for a long time. Bearing these things in mind, I decided to develop a unit capable of reforming and estimating the true capacitance of low-voltage electrolytic capacitors of any value up to about 5000 μF.

Circuit

The basic principle of operation is that a capacitor (C_1) of known value is charged to a known voltage (V_1). The supply is then removed and C_1 is placed across an unknown capacitor C_2. The charge is shared between the two capacitors and the voltage falls to a new lower value (V_2). C_1, V_1 and V_2 are next substituted into the following simple formula to find C_2

$$C_2 = C_1 \frac{V_1 - V_2}{V_2}$$

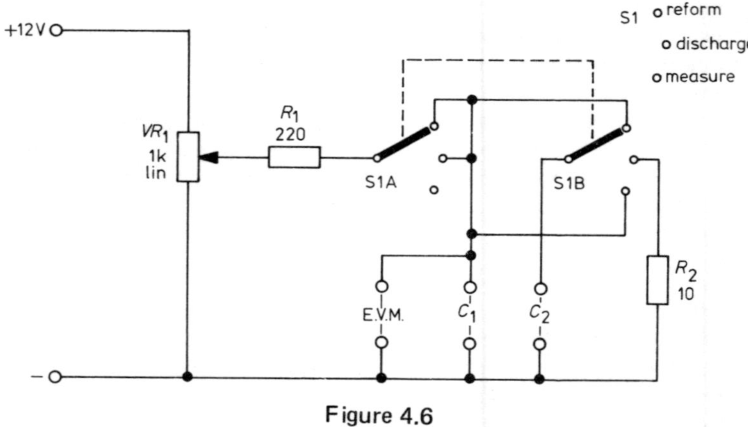

Figure 4.6

In the practical circuit shown in figure 4.6, the three positions of switch S1 select reform, discharge and measure respectively. In the first position, the test capacitor C_2 is reformed. This involves connecting it in parallel with a supply (of correct polarity and voltage set by VR_1) and a high-impedance electronic voltmeter and then waiting for the voltmeter reading to stabilise. VR_1 must be set to the working voltage of C_2 or the reference capacitor (C_1), whichever is the lower. Switching to the second position discharges C_2 through R_2, but ensures the C_1 is fully charged to the voltage (V_1) already set by VR_1. Switching to the third position disconnects the supply and connects C_1 in parallel with the reference capacitor with the result that the voltage across the pair falls to V_2.

For example, if C_1 is 1000 μF, V_1 is 10 V and V_2 is 8 V, the calculation becomes

$$C_2 = C_1 \left(\frac{V_1 - V_2}{V_2}\right)$$

$$= 1000 \left(\frac{10 - 8}{8}\right)$$

$$= 250\ \mu\text{F}$$

If very large capacitors are to be tested, then a very heavy initial charge current will be drawn. I have incorporated R_1 to keep it to reasonable proportions. About the best choice for the reference capacitor is a tantalum component — a very close tolerance device in electrolytic terms. A 100 μF reference capacitor allows easy estimates of test capacitors of up to about 1000 μF. Above this, further tantulums will need to be shunted across the reference capacitor. A break-before-make switch must be used for S1 otherwise the power supply will be shorted as you switch between positions 1 and 2.

Construction

Figure 4.7 shows the front panel layout that I used. I fitted a couple of external terminals for the reference capacitor, two for the test capacitor and two for the electronic voltmeter. I used an external supply for which an extra pair of terminals was provided. I fitted VR_1 and S1 to the opposite end of the chassis and carried out all the wiring in point-to-point fashion. The layout is not at all critical and there is no reason why you shouldn't modify it to suit any container that you have to hand. It may be a good idea to shield the test capacitor because old, leaky ones can sometimes end their lives with a very big bang!

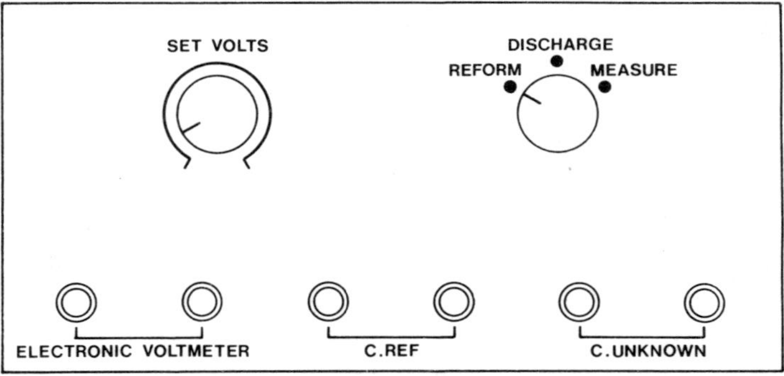

Figure 4.7 Electrolytic capacitor tester, front panel layout

Fault-finding

Assuming that your power pack and electronic voltmeter are all right, then the only problem with this circuit is possible confusion over the switch wiring. If your circuit fails to operate, check the switch wiring again. If the test capacitor is an old one, its leads may be oxidised and not making good contact.

Simple Diode Tester

A very cheap way of obtaining diodes is to buy bumper packs of unmarked, untested devices. The circuit in figure 4.8 can be used to reject quickly any components that are short circuit or open circuit. It can then be used to determine the polarity of all the good diodes.

Connect your test diode across terminals A and B. If either lamp 1 or lamp 2 lights, the diode is a good one. If both lamps light, it is short circuit and if it is open circuit, neither lamp will light. If the cathode of a good test diode is connected to terminal B as shown, then lamp 1 will light. To check the lamps, just short terminals A and B together, when both lamps should light.

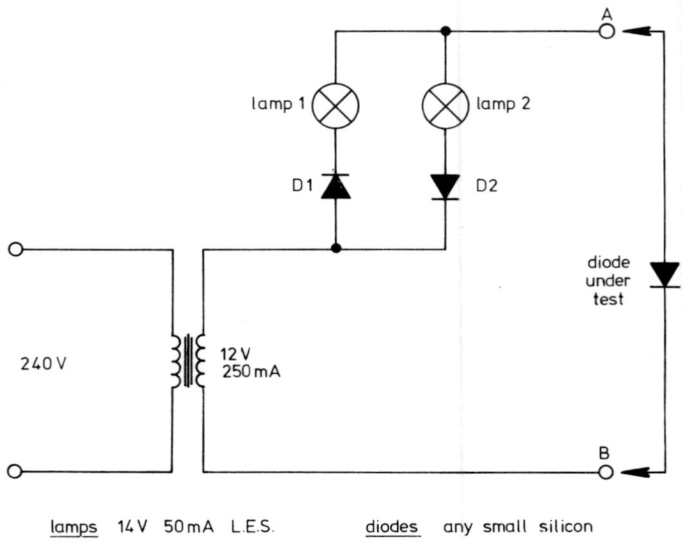

Figure 4.8

Transistor Tester

Very many projects for the home constructor do not really necessitate using a specific type of transistor – often almost any general-purpose device will do. Perfectly serviceable components can be recovered from old circuit boards, but these need to be checked before being re-used. The circuit suggested here will enable you to

If the wiring is correct but you get the wrong output on ac only, then the oscillator section is faulty. If you get the wrong outputs on both dc and ac, then the fault probably lies between the input of TR3 and the output terminals.

Having isolated the faulty sections you must change components in the order of semiconductors, capacitors and lastly resistors, checking the circuit function after each stage.

Oscilloscope Dual Trace Adaptor

If you have a single trace oscilloscope, you will know how useful it is for measuring voltage, current and inspecting waveforms. A dual trace oscilloscope is even more useful, but unfortunately it is much more expensive. The adaptor suggested here (see figure 5.3) extends most of the advantages of the dual trace facility to the owner of a cheaper single trace model. Namely, it enables you to compare two different signals for voltage, waveform, phase and so on.

The device will handle input signal frequencies of more than 1 MHz but, on the assumption that it will be used with semiconductor circuitry, I have only allowed for a maximum input voltage of about 5 V r.m.s. Each channel has an input impedance of approximately 100 kΩ but both this and the input voltage limit may be raised by means of a simple external attenuator – a suitable one is shown in figure 5.8. Continuously variable gain controls are fitted to each channel and although I did not calibrate mine, this should not be too difficult to do using sources of known voltage.

Circuit

The complete circuit is shown in figure 5.4. The input sockets for channel A and channel B drive TR1 and TR2 respectively. These transistors, connected as bootstrapped emitter followers to give a high input impedance, are used to drive the diode gating circuit formed by D3 and D4. VR_1 and VR_2 are continuously variable gain controls for each channel. TR1 and TR2 emitters are also connected to the sync output sockets so that the oscilloscope may be synchronised by either the channel A signal or the channel B signal. TR3 and TR4 form the chopper oscillator, which may be run at one of two speeds. D1 and D2 improve oscillator stability, R_{12} and D5 improve the risetime of the output waveform. The chopper signal is direct coupled into the base of the buffer amplifier, TR5. VR_3 in TR5 collector determines the amplitude of the chopping signal fed to the diode gating circuit and thus provides control of trace separation.

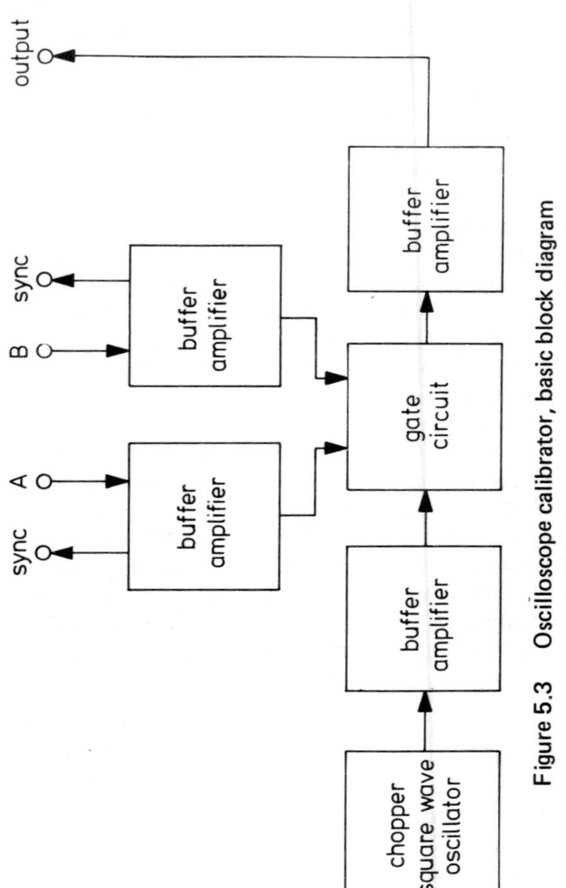

Figure 5.3 Oscilloscope calibrator, basic block diagram

+18 V
S1
sync. A
2
input A
1
input B
5
sync. B
4
−
C_1 50 µ
C_2 12 µ
C_3 220n
C_4 10 µ
C_5 220n
C_6 10 µ
C_7 12 µ
C_8 1n
C_9 5n
C_{10} 1n
C_{11} 5n
C_{12} 47n
C_{13} 12 µ
C_{14} 10 µ
C_{15} 12 µ
C_{16} 12 µ
C_{17} 12 µ
R_1 470
R_2 47k
R_3 22k
R_4 56k
R_5 22k
R_6 47k
R_7 56k
R_8 470
R_9 4.7k
R_{10} 82k
R_{11} 82k
R_{12} 4.7k
R_{14} 4.7k
R_{15} 680
R_{16} 18k
R_{17} 8.2k
R_{18} 8.2k
R_{19} 120k
R_{20} 82k
R_{21} 5.6k
VR_1 4.7k lin
VR_2 4.7k lin
VR_3 4.7k lin
t
w
b
TR1 2N3904
TR2 2N3904
TR3 2N3904
TR4 2N3904
TR5 2N3904
TR6 2N3904
D1 1N914
D2 1N914
D3 1N914
D4 1N914
D5 1N914
S2A
S2B
i
ii
out
3
c
b
e
2N3904 from wiring side
a
k
1N914

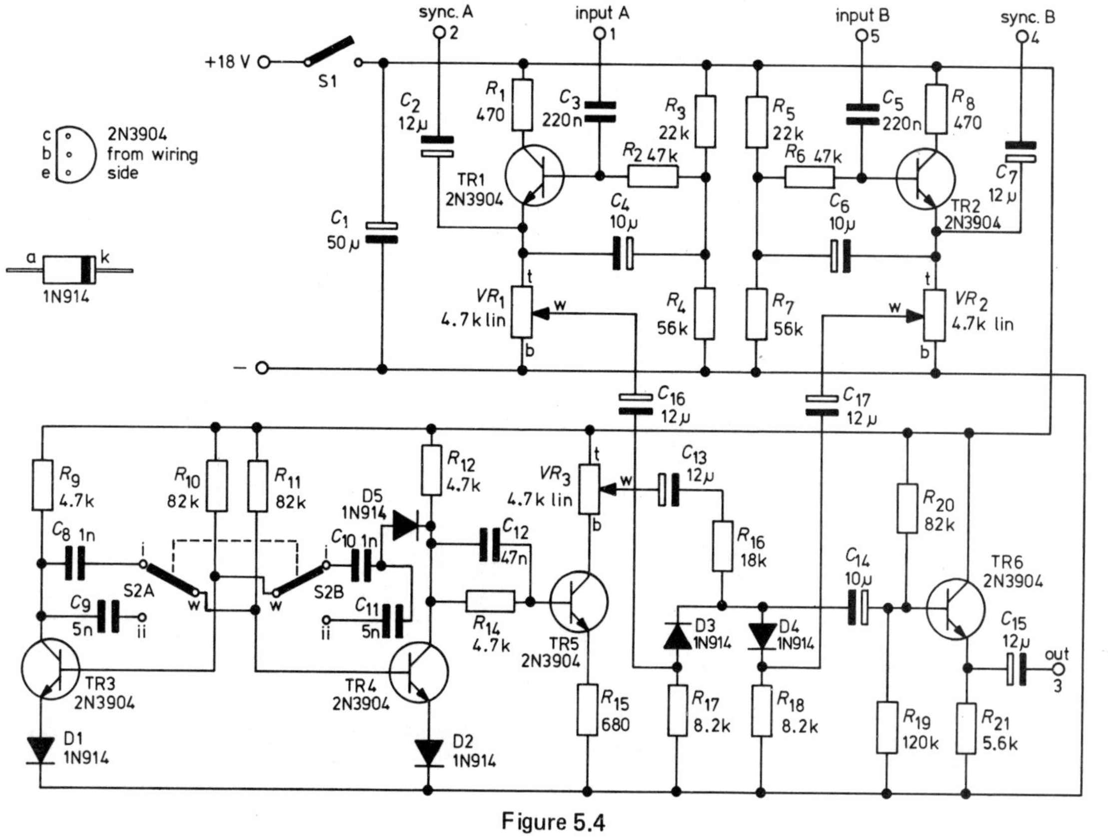

Figure 5.4

This signal switches rapidly between inputs A and B, and results in the appearance of two separate traces on the oscilloscope screen that reproduce the signals applied to the respective inputs. TR6 is a buffer amplifier providing plenty of low impedance drive to the oscilloscope.

Construction

I fitted all the switches, sockets and variable controls to a 202 mm x 102 mm front panel, and, as I used an external 18 V supply, this was screwed to a 77 mm deep box. A deeper one will be needed if you are going to use batteries. I fitted the rest of the components – most of them vertically to save space – to a single piece of Veroboard. Remember that fairly high frequencies are being handled and this makes the layout quite critical. Follow the diagram shown in figure 5.5 and keep all interconnecting leads and links as short as possible. Fit all the interconnecting leads and links first, followed by the resistors and capacitors. Lastly, fit the semiconductors. A suitable front panel layout is shown in figure 5.6. This shows provision for co-axial sockets as these fitted into my system. Most modern oscilloscopes have BNC sockets and you may wish to use these – making holes in the front panel to suit. The over-all layout that I used is shown in figure 5.7.

Testing and Fault-finding

Adjust your oscilloscope *Y* amplifier sensitivity to approximately 100 mV/cm. Switch on the dual trace adaptor and connect its output (socket 3) to the *Y* input socket of the oscilloscope. Using internal sync, make sure that the adaptor produces a good squarewave on both high and low chopping speeds – you will have to adjust the oscilloscope timebase to do this. You should also be able to vary the height of the squarewave using the trace separation control. Next, reduce the timebase speed until two horizontal traces are obtained. Connect test inputs to channel A and B input sockets (sockets 1 and 5 respectively). If your oscilloscope has an external sync facility, switch to this and connect the appropriate socket to socket 2 or socket 4 of the dual trace adaptor. You should now have two test traces on the screen. The gain control for each channel should vary the appropriate trace and you should be able to control the distance between the traces using the trace separation control. Don't attempt to superimpose the two traces because this will cause them to interact and break up. If any fault-finding is necessary, check your wiring first. When there are so many of them, it is very easy to leave out a vital link or even a component. If all is correct, use your oscilloscope to

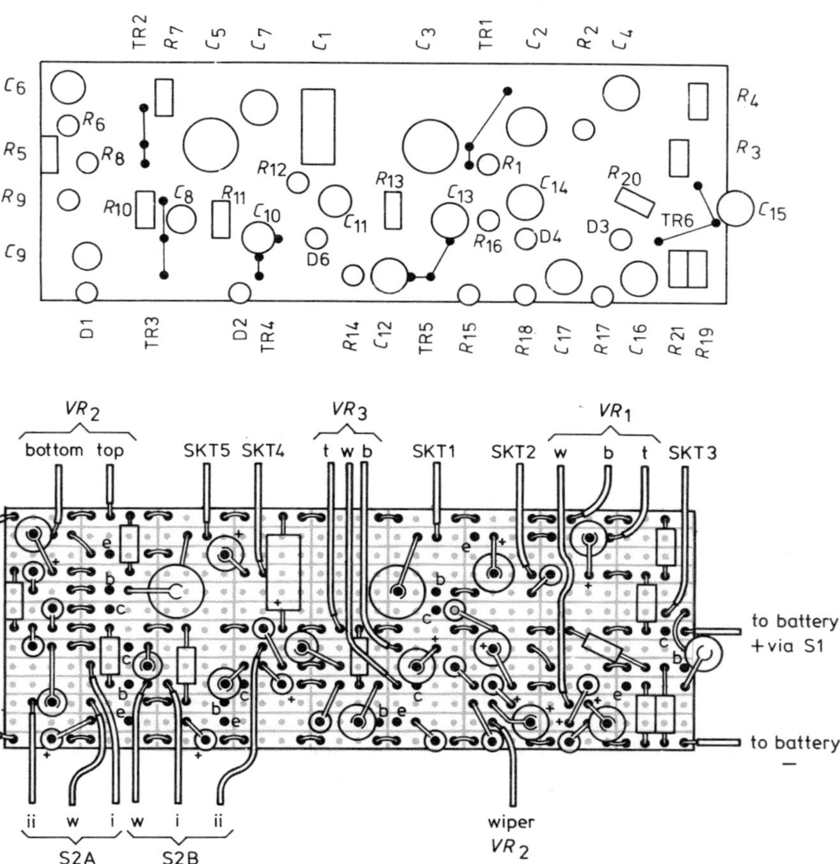

Figure 5.5

trace the squarewave of the chopper oscillator to find out where it gets lost — if indeed the chopper is working properly at all.

When you are using the unit, you will find that there is noticeable interaction between the individual channel gain controls and the trace separation control. There is also a certain amount of crosstalk or breakthrough between the channels. However, in practice I don't think this will worry most users.

A simple way of increasing the input impedance and signal voltage capability is to make up an oscilloscope high-impedance probe. The

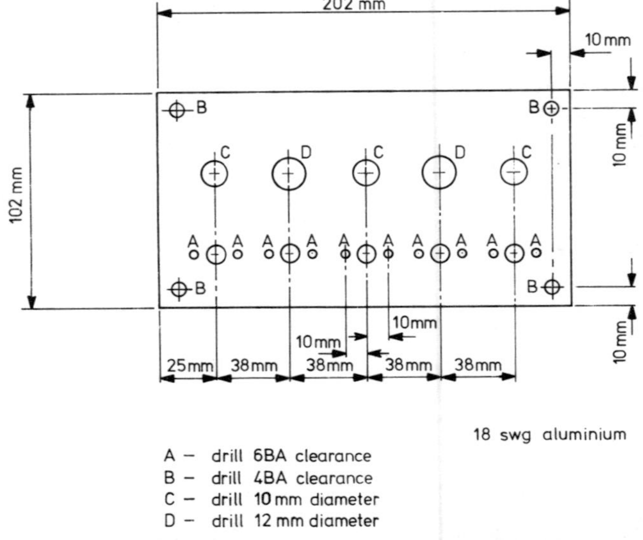

Figure 5.6 Oscilloscope dual trace adaptor, front panel layout

one shown in figure 5.8 increases the input impedance to about 1 MΩ and the input voltage upper limit is raised to about 50 V r.m.s. The trimmer capacitor is in series with the input stray capacitance and so reduces it. Adjust the trimmer to obtain the best possible squarewave.

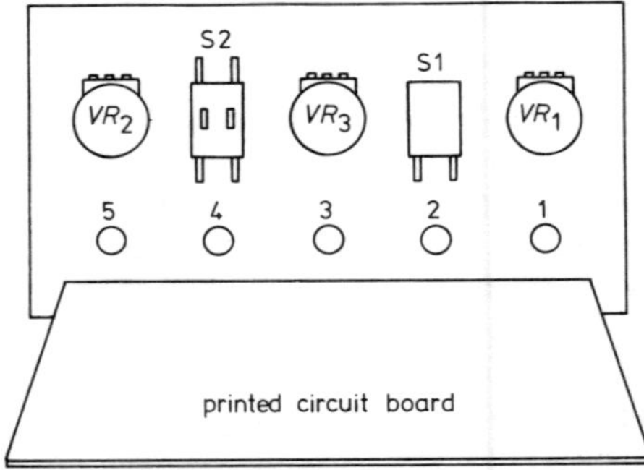

Figure 5.7 Oscilloscope dual trace adaptor, front panel drilling details

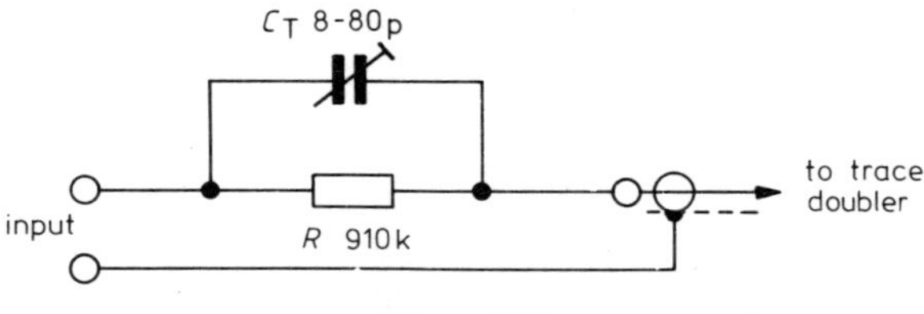

Figure 5.8

A Simple Sawtooth Generator

This simple sawtooth generator produces a linear output with a frequency variable between a few Hz and a few hundred Hz. Among other applications, the generator can be used as a timebase oscillator or as part of a voltage-controlled oscillator.

Circuit

The first transistor (TR1), in the circuit shown in figure 5.9, is a constant current generator. C_1 charges up through this until it reaches about 80 per cent of the supply rail voltage. At this point

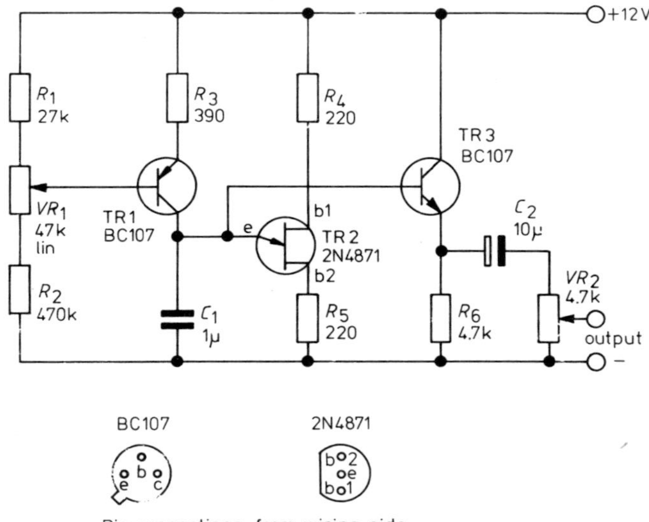

Figure 5.9

the unijunction transistor (TR2) conducts, quickly discharging the capacitor. The sawtooth waveform produced at the emitter of TR2 is taken to the output level control (VR_2) via the emitter follower buffer amplifier TR3.

The frequency of the sawtooth is determined by the setting of VR_1, which determines the charge rate of C_1. If required, a positive-going needle pulse waveform is available at B1 of TR2 and a negative-going needle pulse may be obtained from B2 of TR2.

6
Workshop Gadgets

Soldering-iron Saver

Like a watched kettle, a soldering iron seems to take ages to warm up when it is first switched on. If you avoid this frustration by leaving your iron switched on all the time, even an iron-coated bit will have to be replaced in an alarmingly short time. The scheme I have shown in figure 6.1 allows you to leave the iron on standby when you are not actually using it. You need a switched plug and

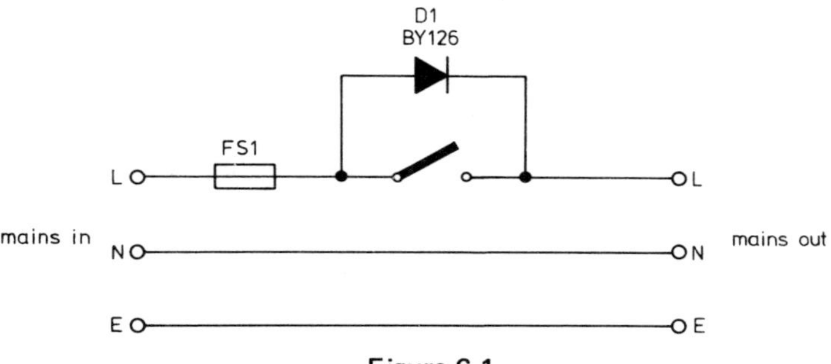

Figure 6.1

the rectifier diode is connected – either way round – across the switch. You must, of course, choose a plug where you have access to the switch.

A Low-noise Switch

If you need to switch the sensitive input circuitry of an a.f. or r.f. amplifier, switching clicks and noise pick up can be a problem. In the circuit shown in figure 6.2, an FET is used to avoid these problems. An *n*-channel JUGFET is a normally on device and, with S1 in the 'off' position, the FET is conducting hard and will short out the incoming signal. With S1 in the 'on' position the FET is biased beyond cut off and its high impedance will not substantially interfere with the signal. S1 can be remotely mounted, but you can fit the FET near the input circuit to avoid pick up of hum and noise.

High Input Impedance Buffer Amplifier

Although the bipolar transistor is an immensely useful device, it does have the disadvantage that it puts a relatively heavy load on the cir-

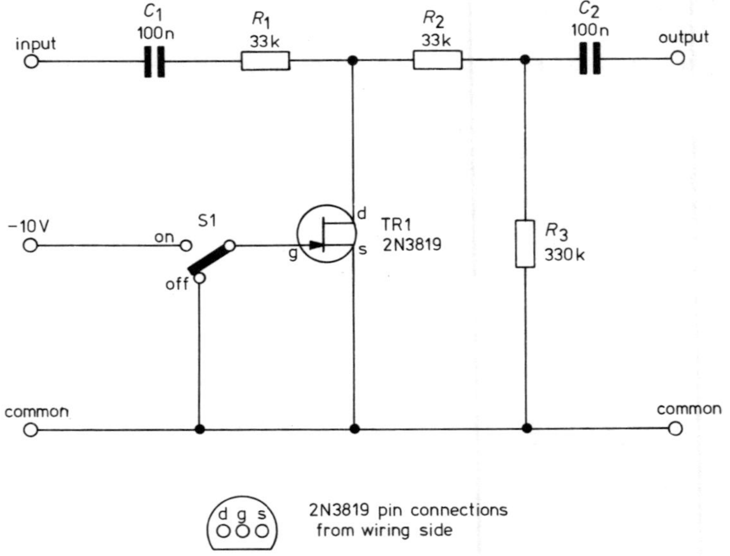

Figure 6.2

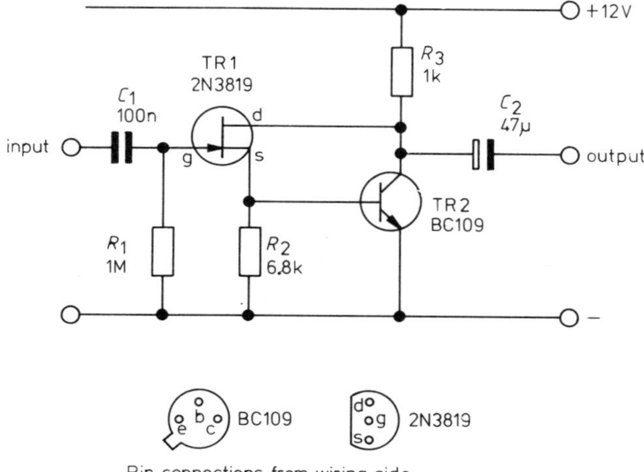

Figure 6.3

cuit that it is driven by. The circuit I have suggested in figure 6.3 uses the very high input impedance of an FET at its front end. It has a voltage gain of about 50 V, and input impedance of about 750 kΩ and will handle input signals of up to about 100 mV peak to peak.

A Bench Test Audio Amplifier

There are many occasions when an audio amplifier would be useful in the workshop. You can use this one to amplify oscillator signals, or indeed bring any small signal up to a level where it can be used to drive a heavy load like a relay or loudspeaker.

Circuit

The circuit, using five small transistors, is shown in figure 6.4. VR_1 is the input gain control and the signal is coupled from its wiper to the base of TR1 by means of C_1. The amplified signal is direct coupled from the collector of TR1 to the base of the driver amplifier, TR2. This drives the complementary output pair TR3 and TR4, which – since they are connected as emitter followers – have a voltage gain of less than 1. C_6 connects an in-phase signal to the top of R8 thus providing positive feedback, increasing the voltage gain of TR2,

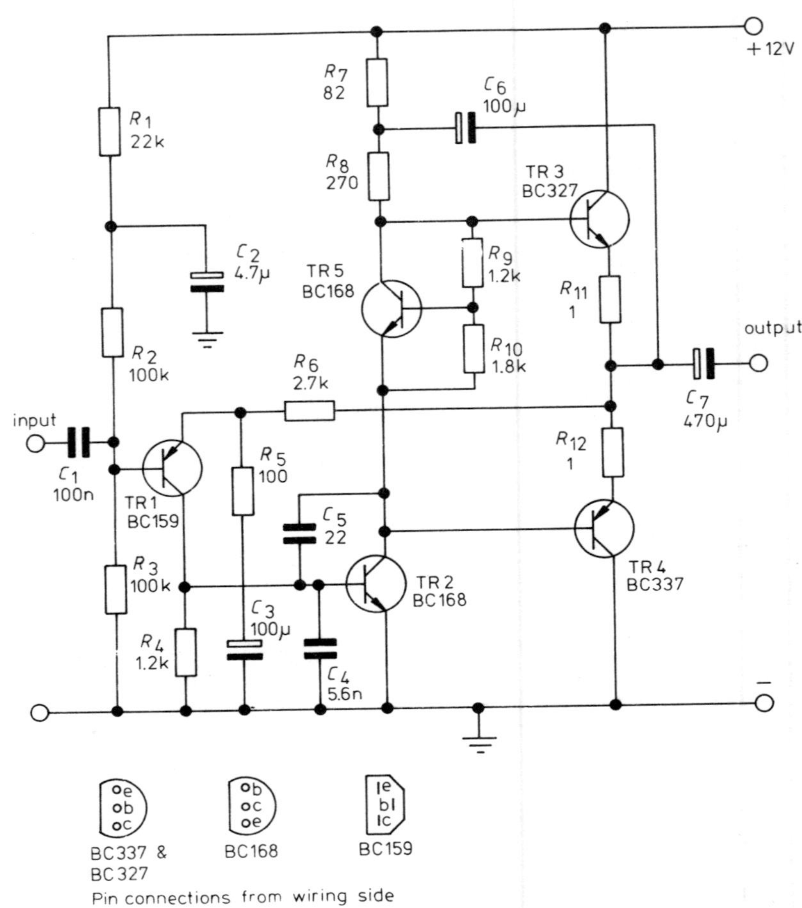

Figure 6.4

ensuring that it can provide a really good voltage drive to TR3 and TR4. This technique is known as 'bootstrapping'.

Transistor TR5 is used to set the bias voltage for the output transistors, turning them on slightly, thus reducing cross-over distortion. R_6 and R_5 provide negative feedback, but do not effect output stage biassing as R_5 is isolated by means of C_3. Modern silicon transistors like these will amplify signals right up to radio frequencies but capacitors C_4 and C_5 are fitted to make sure that they don't get the chance!

I have specified a 12 V supply rail but, provided that you can accept reduced performance, this can be reduced or increased if you wish. Current requirements are up to about 750 mA.

Construction

Fit all the components on to a small piece of Veroboard – I have shown a suitable layout in figure 6.5. The heatsink is about 25 mm x 25 mm x 25 mm high and I attached TR3, TR4 and TR5 directly to it, using fast-acting epoxy resin. For this reason, it is important that you use the specified transistors in their TO 92 plastic package form. If you use other types, the collector is usually common to the metal case and the transistors will be shorted out by the heatsink. If you are going to use the amplifier with a loudspeaker, it is this component that will determine the size of any container that you use. About a 100 mm speaker is suitable. I used a built-in loudspeaker but decided to use an external power supply. An aluminium box 115 mm x 250 mm x 75 mm deep will provide enough space for the circuit board, a loudspeaker and a power supply as well if you want one.

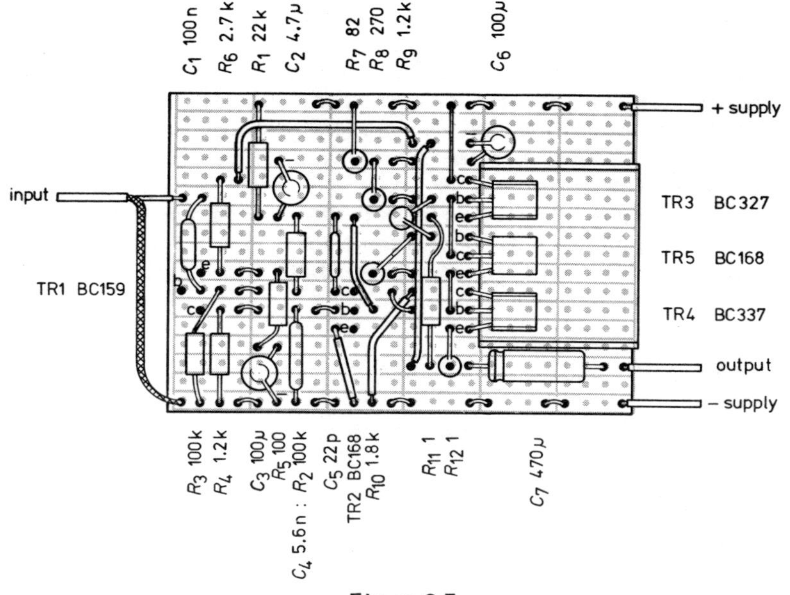

Figure 6.5

Testing

When you have completed the construction and carefully checked over the wiring, measure the resistance across the supply rail. This should be low initially, but rise as the large decoupling capacitors charge up. Connect a 1 kΩ $\frac{1}{2}$ W resistor across the output and a 12 V power pack to the supply rail. When you switch on, you should find that the unit draws just a few mA under no signal conditions. Using a high-resistance voltmeter, check that you have about $V_{cc}/2$ at the junction of R_{11} and R_{12} with VR_1 at its highest setting. Inject an a.f. signal of about 1 kHz at 100 mV into the input. You should get a good sinewave output on an oscilloscope. This should be maintained throughout the audio spectrum.

Fault-finding

If the amplifier draws a very heavy current when you switch on, switch off! This is probably caused by a low-resistance path across the supply rails (for example, check C_2 and C_6 for short circuit).

A distorted output for the correct input denotes a possible bias fault or faulty coupling capacitor. If the output transistors are running hot, then C_7 is short circuit or one of the resistors R_7 to R_{10} inclusive may have changed in value or be open circuit.

High noise levels at the output may be caused by a noisy transistor or resistor. To locate the faulty part, try holding a warm soldering iron near each resistor and transistor in turn. If the noise level rises as the iron is brought near any component then that component should be changed for a new one. Such a test is not entirely reliable and a more professional method would be to use local cooling rather than heating. A freezer spray, like the R.S. components one, is useful for this.

No output at all denotes a break in the signal path or that the signal is being shorted out. Check through the circuit with your oscilloscope in order to locate the stage at which the signal is lost.

r.f. Probe II

This probe can be used in conjunction with an a.f. amplifier to trace r.f. signals when you are fault-locating.

The circuit is shown in figure 6.6. D1 demodulates the signal and TR1 amplifies the a.f. and provides enough output to drive an audio amplifier. If you mount all the components on to a small piece of Veroboard, you can build it into a cigar tube type container with the probe tip insulated from the metal casing. Fit a twin-screened lead to

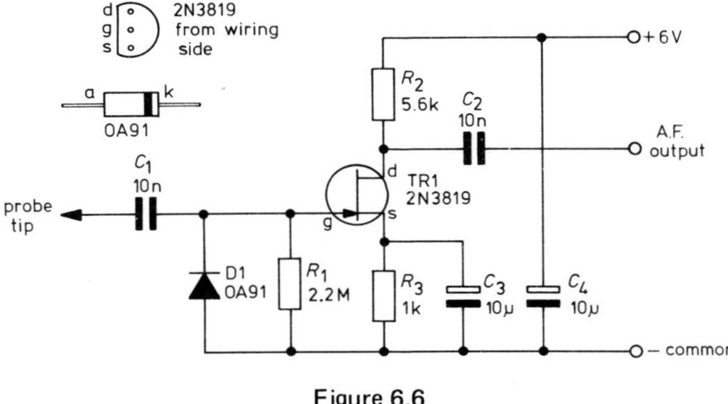

Figure 6.6

take the signal to the amplifier and pick up the required 6 V power supply.

Constant Impedance Attenuator

The function of an attenuator is to reduce any available signal to the required level. In some cases – TV or radio aerial inputs, for example – it is necessary to maintain a constant input and output impedance. Either of the simple circuits shown in figures 6.7 and 6.8 can do this.

The T Circuit

$$R_1 = R\left(\frac{A-1}{A+1}\right)$$

and $$R_2 = R\left(\frac{2A}{A^2-1}\right)$$

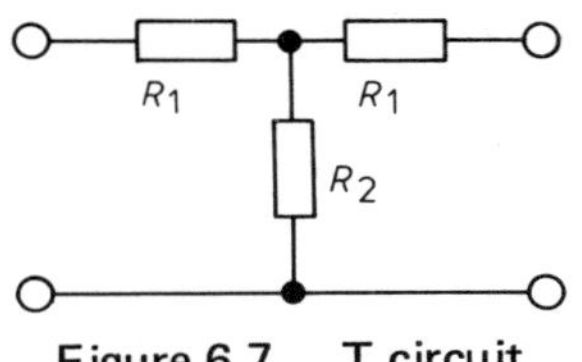

Figure 6.7 T circuit

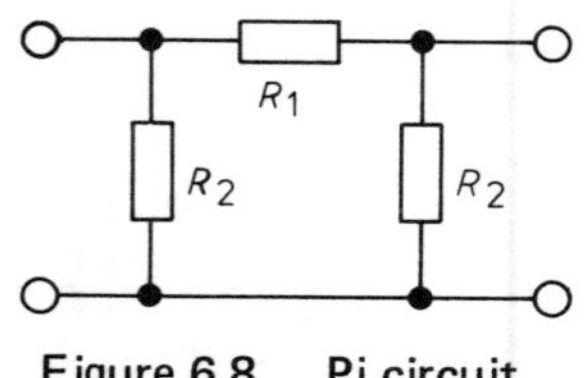

Figure 6.8 Pi circuit

The Pi Circuit

$$R_1 = R\left(\frac{A^2 - 1}{2A}\right)$$

and $$R_2 = R\left(\frac{A + 1}{A - 1}\right)$$

where R is the required impedance and A is the ratio of input to output.

Example Suppose $R = 80\ \Omega$ and a loss of 20 dB (that is 10 to 1) is required, then

(a) For T circuit

$$R_1 = 80\left(\frac{10 - 1}{10 + 1}\right) = 65.45\ \Omega$$

and

$$R_2 = 80\left(\frac{2 \times 10}{10^2 - 1}\right) = 16.2\ \Omega$$

(b) For Pi circuit

$$R_1 = 80\left(\frac{10^2 - 1}{2 \times 10}\right) = 396\ \Omega$$

and

$$R_2 = 80\left(\frac{10 + 1}{10 - 1}\right) = 97.8\ \Omega$$

In this example the values required for the Pi type are more readily available although this is not always the case. You must calculate which would be the more suitable circuit for your needs.

A Simple Timer Circuit

This circuit provides timing of periods from a few milliseconds to 8 minutes. It indicates that the pre-set time has elapsed by means of an audible warning from a loudspeaker.

Circuit

The circuit is shown in figure 6.9. With S1 in the position shown C_1 charges up to the full battery voltage. IC1 will oscillate if more than two-thirds of the supply voltage is applied to pins 2 and 6. The loudspeaker is connected to its output (pin 3) so that when it oscillates a signal can be heard.

The voltage at pins 2 and 6 is normally held low, but when S1 is switched over, C_1 discharges through TR1 base–emitter junction.

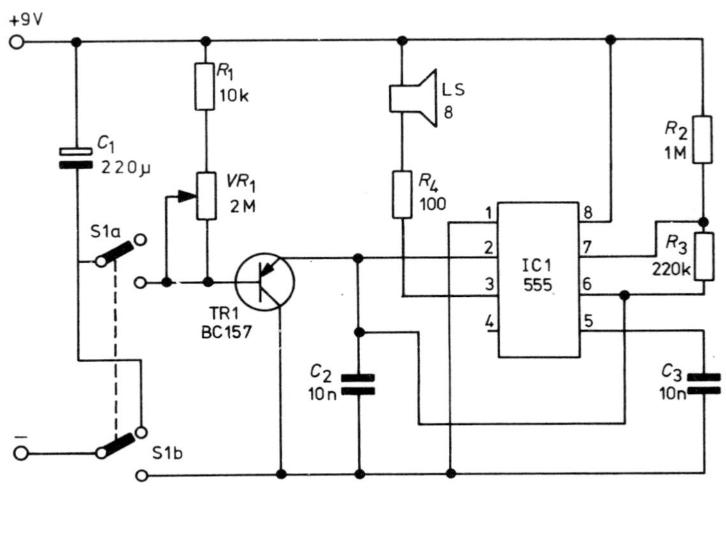

Figure 6.9

The rate of discharge is set by the bias components R_1 and VR_1. The voltage at pins 2 and 6 slowly rises until it reaches two-thirds of supply where IC1 starts to oscillate. Re-setting S1 starts the cycle again.

Construction

Fit the components on to a small piece of Veroboard as shown in figure 6.10 using an 8-pin DIL socket for IC1. Mount this, the loudspeaker, VR_1 and the battery into a small box – about 150 mm x 75 mm is adequate. You can calibrate VR_1 later using Letraset to mark the front panel. Be sure to fit IC1 and C_1 the right way round.

Testing

When you have completed the construction and carefully checked your wiring, fit a known good PP3 battery with S1 in the off position as shown in figure 6.9. Turn VR_1 to maximum resistance and switch S1. About 7 to 8 min should elapse before the circuit starts to whistle. Re-set S1 and turn VR_1 to minimum resistance. The circuit should oscillate a few seconds after switching S1 on again.

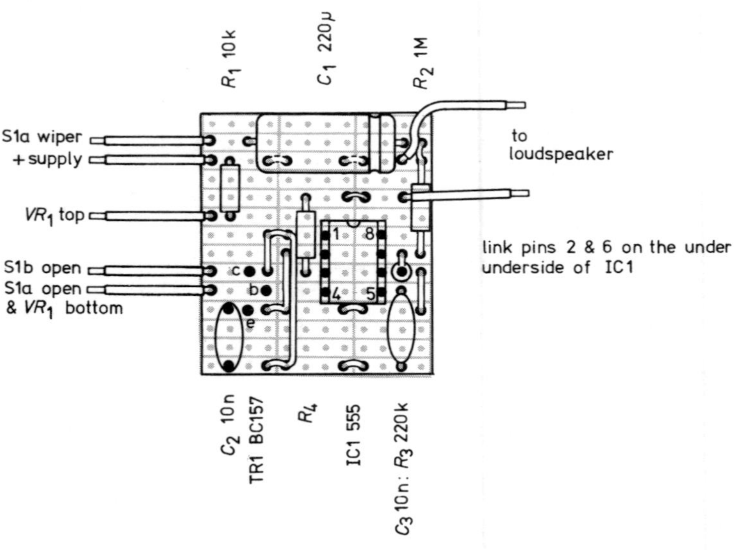

Figure 6.10

Fault-finding
If the circuit fails to operate, check first whether you are using a good battery. If it is all right check the loudspeaker by removing it from the circuit and connecting an ohmmeter across it. It should click as the meter is connected. If this is working, refit the speaker, lift off emitter connection and switch S1 to the on position. If IC1 does not oscillate, then it must be faulty – replace it and try again. If connecting TR1 stops IC1 oscillating or if the elapsed time before oscillation starts is not influenced by VR_1, then TR1 is probably faulty.

Resistance Substitution Box

One of the biggest timewasters, when you are trying to develop a piece of project work, is having to hunt through your spares box for a particular value of resistor – and then not finding one! This project will provide you with an example of most standard values of resistance at your finger tips.

As well as being useful when developing new circuits, this resistance substitution box is invaluable for servicing work. You can quickly find the value of a burnt-out component, for example.

Circuit
The circuit, shown in figure 6.11, is quite straightforward. I used two 12-position single-pole rotary switches for S1 and S2. You could, of course, use a 12-position two-pole switch if you can get one. In my design, S1 selects any one of 11 resistors between 10 Ω and 10 kΩ. S2 selects any one of 11 resistors between 15 kΩ and 10 MΩ. The remaining position of each switch gives 'off'. S3, which can be any single-pole changeover switch, selects either low values (S1) or high values (S2). The resistors – R_1 to R_2 inclusive – are all $\frac{1}{2}$ W 5 per cent tolerance carbon film components.

Construction
The construction that I used is shown in figure 6.12. I fitted all the resistors directly to the appropriate switch contacts using a loop of heavy gauge tinned copper – about 18 s.w.g. will do – for the common side of the resistors. I didn't think it was worth fitting output terminals. I used a pair of PVC-covered flexible leads fitted with insulated croc-clips as I thought these would cope with all situations. For the front panel signwriting I used Letraset dryprint and in order to avoid the problem of the Ω sign, I used the BS 1852 code. In this coding, which is being widely adopted by commercial manufacturers,

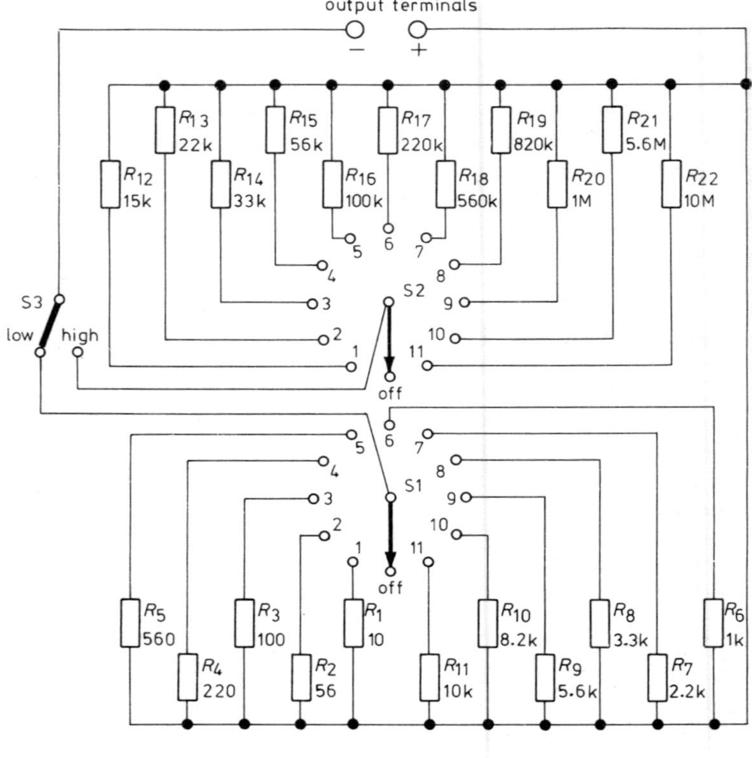

Figure 6.11

R is used instead of Ω, k instead of kΩ and M instead of MΩ, these letters replacing the decimal point. For example

4R7 signifies 4.7 Ω

4k7 signifies 4.7 kΩ

1M0 signifies 1 MΩ

56k signifies 56 kΩ

and so on

The unit was fitted into an aluminium box about 100 mm x 60 mm x 50 mm deep.

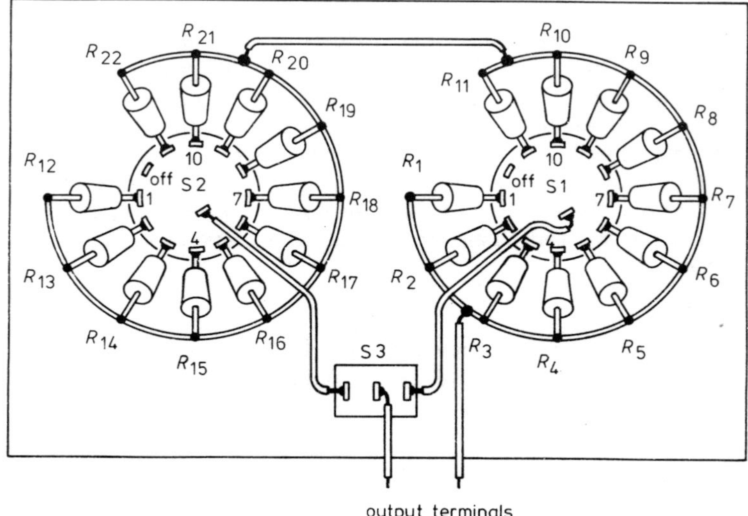

Figure 6.12 Resistance substitution box, suggested layout

Testing and Fault-finding

When you have constructed and carefully checked your substitution box you should ensure that each switch position selects the appropriate resistor. You can do this using an ohmmeter or resistance bridge.

If any switch position indicates the wrong resistance – outside normal tolerance limits – then that resistor should be changed. If any position gives no resistance then the switch contact is probably faulty. Clean it using a little carbon tetrachloride applied with a small brush and turning the switch back and forth a few times.

Capacitor Substitution Box

Have you ever had to postpone finishing a project for the want of a particular value capacitor? Wouldn't it be useful if you always had the right value handy to bridge across, or replace, that suspect capacitor in the unit you are servicing?

This project provides you with a wide selection of capacitors, instantly available at the turn of a switch.

Circuit

My circuit is shown in figure 6.13. I used a pair of 12-position, single-pole rotary switches for S1 and S2. S1 selects one of 11 capacitors between 2.2 pf and 100 nF. S2 selects one of the remaining capacitors between 1 μF and 1000 μF. The choice of the larger components depends on their physical size. I suggest you use the highest working voltage, for the given value of capacitance, that will fit inside your chosen box. S3 is a single-pole changeover switch, which selects the low capacitance (S1) range or the high capacitance (S2) range.

Construction

I used the construction shown in figure 6.14. Connect the capacitors directly to their appropriate switch contacts and connect the 'com-

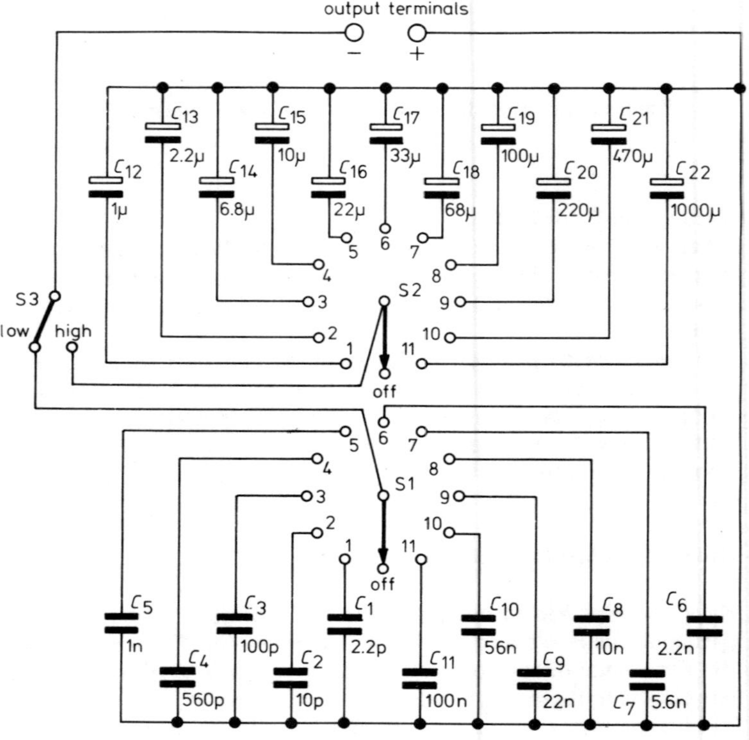

Figure 6.13

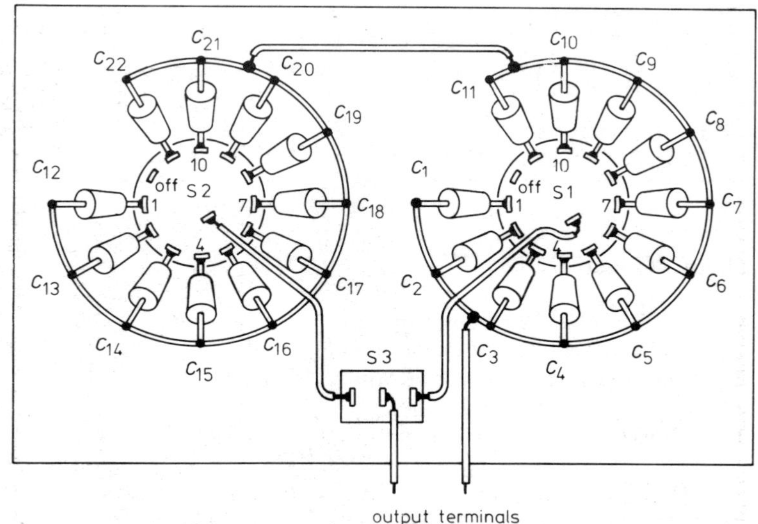

Figure 6.14 Capacitor substitution box, suggested layout

mon' ends together using a loop of tinned copper wire as I have shown. Cover the capacitor leads with PVC sleeving where there is any possibility of them touching any other metal parts. I used a pair of flexible leads fitted with insulated croc-clips for the output connections. I marked my front panel with Letraset dryprint and to simplify this job I used the coding recommended by BS 1852. In this code, μ replaces μF, n replaces nF and p replaces pF, these symbols replacing the decimal point. For example

2μ2	means	2.2 μF
100μ	means	100 μF
4n7	means	4.7 nF
2p2	means	2.2 pF

and so on

If you have trouble with the μ sign, you can use a letter '*u*' adding a small tail to the front 'μ'.

I fitted the whole unit into a small aluminium box about 100 mm x 60 mm x 50 mm deep.

Appendix I: Components required for Projects

Simple Regulator

R_1 2.2 kΩ ±10% $\frac{1}{4}$ W carbon film
C_1 10 nF polyester 250 V dc working
C_2 100 nF polyester 250 V dc working
Z1 BZX83 13 V 400 mW Zener
TR1 BC107 (or similar) silicon *npn*

Regulated Power Supply for CMOS and 74 Series ICs

R_1 1 kΩ ±10% $\frac{1}{4}$ W carbon film
R_2 3.3 kΩ ±10% $\frac{1}{4}$ W carbon film
R_3 3.3 kΩ ±10% $\frac{1}{4}$ W carbon film
VR_1 5 kΩ linear potentiometer carbon track
C_1 2200 μF 40 V dc working electrolytic
C_2 330 nF polyester 250 V dc working
C_3 100 nF polyester 250 V dc working
D1 WOO5 1 A 50 V bridge rectifier
Reg1 μA78MO5UC 500 mA 5 V regulator or similar
IC1 μA741C op-amp
T1 mains transformer with 9-0-9 V secondary rated at 250 mA minimum
FS1 100 mA 20 mm type
S1 S.P.S.T. miniature toggle switch
S2 D.P.D.T. miniature toggle switch

Simple Dual Polarity Power Supply

R_1 20 Ω ±10% $\frac{1}{2}$ W carbon film
R_2 560 Ω 5 W wire wound
C_1 68 μF 63 V dc working electrolytic
D1–D4 IN4002 100 V 1 A silicon diode
Z1and Z2 IN4744A 15 V 1 W Zener

Balanced Supply From One Winding

R_1 220 Ω ±10% 1 W carbon film
C_1 2200 μF 35 V dc working electrolytic
C_3 220 μF 35 V dc working electrolytic
D1–D6 IN4002 1 A 100 V silicon diode
T1 mains transformer with 18 V 1 A secondary

200 V from a PP3!

R_1 10 kΩ ±10% $\frac{1}{2}$ W carbon film
C_1 100 μF 16 V dc working electrolytic
C_2, C_3 47 nF polyester 250 V dc working
C_4, C_5 470 nF metallised polyester 630 V dc working
D1, D2 BY127 1 A 1200 V silicon diode
TR1 BC107 (or similar) silicon *npn*
T1 240 V to 9-0-9 V mains transformer

Blown Fuse Indicator II

R_1 1 MΩ ±10% $\frac{1}{2}$ W carbon film
R_2 100 kΩ ±10% $\frac{1}{2}$ W carbon film
C_1 220 nF polyester 250 V dc working
D1, D2 IN4005 1 A 600 V silicon diode

Power Supply Protection with a Crowbar

R_1 100 Ω ±10% $\frac{1}{2}$ W carbon film
R_2 10 Ω ±10% $\frac{1}{2}$ W carbon film
C_1 0.68 μF 10 V dc working electrolytic
Z1 BZX83 6.2 V 400 mW Zener
SCR TIC 106A 4 A 100 V thyristor

Two-transistor Signal Injector

R_1, R_4 10 kΩ ±10% $\frac{1}{4}$ W carbon film
R_2, R_3 330 kΩ ±10% $\frac{1}{4}$ W carbon film
VR_1 10 kΩ linear potentiometer carbon track
C_1, C_2 10 nF polycarbonate 100 V dc working

C_3 10 nF polyester 500 V dc working
TR1, TR2 BC107 (or similar) silicon *npn*
S1 push-to-make miniature push button

Logic Probe for TTL

R_1 56 kΩ ±10% $\frac{1}{4}$ W carbon film
R_2 560 Ω ±10% $\frac{1}{4}$ W carbon film
R_3, R_4 100 Ω ±10% $\frac{1}{4}$ W carbon film
D1, D2 TIL 209 red LED
IC 7404 hex inverter
TR1 BC107 (or similar) silicon *npn*

a.f. Signal Generator

All resistors ±10% $\frac{1}{4}$ W carbon film types unless otherwise stated
R_1, R_2 1 kΩ
R_3 type R53 glass bead thermistor (R.S. Components)
R_4 470 Ω
R_5, R_6 3.3 kΩ
VR_1 10 kΩ linear dual gauged potentiometer carbon track
VR_2 5 kΩ linear potentiometer carbon track
C_1, C_4 680 nF polyester 250 V dc working
C_2, C_5 68 nF polyester 250 V dc working
C_3, C_6 6.8 nF ceramic 63 V dc working
C_7, C_8 47μF 16 V dc working electrolytic
IC1 μA741 op-amp
S1 2-pole 3-position rotary switch
S2 S.P.S.T. miniature toggle switch

Crystal Reference Oscillator

R_1, R_2 470 Ω ±10% $\frac{1}{4}$ W carbon film
C_1 100 nF polyester 250 V dc working
CT1 5.5 pF to 65 pF miniature film dielectric trimmer
X1 HC-6/u 1 MHz plug-in crystal
IC1 7400 quad 2-input positive NAND gate
IC2—IC7 7490 decade counter

Waveform Generator

All resistors are ±10% $\frac{1}{4}$ W carbon film unless otherwise stated

R_1 18 kΩ
R_2 10 MΩ ±5% $\frac{1}{2}$ W
R_3, R_4 4.7 kΩ
R_5, R_7 15 kΩ
R_6 47 kΩ
R_8, R_9 10 kΩ
VR_1, VR_3 1 kΩ linear carbon preset potentiometer
VR_2 10 kΩ linear carbon potentiometer
VR_4, VR_5 100 kΩ linear carbon preset potentiometer
VR_6 100 kΩ linear carbon potentiometer
VR_7 10 kΩ linear carbon preset potentiometer
C_1 100 nF polyester 250 V dc working
C_2 4.7 nF silvered mica 350 V dc working
D1 IN 914 switching diode
IC1 ICL 8038 waveform generator
IC2 μA741c op-amp
S1 single-pole 3-position rotary switch
S2 D.P.D.T. miniature toggle switch

White Noise Generator

R_1 5.6 kΩ ±10% $\frac{1}{4}$ W carbon film
R_2 56 kΩ ±10%-$\frac{1}{4}$ W carbon film
C_1 20 μF or 22 μF 40 V dc working electrolytic
C_2 1 μF or 22 μF 40 V dc working electrolytic
TR1, TR2 BC107 (or similar) *npn* silicon

Stereo Balance

All resistors ±10% $\frac{1}{4}$ W carbon film unless otherwise state
R_1, R_4 15 kΩ
R_2, R_3 330 kΩ
R_5, R_8 390 Ω
R_6, R_7 2.2 kΩ
R_9 60 Ω ±5% 5 W wire-wound
VR_1 10 kΩ linear carbon potentiometer
C_1–C_3 10 nF polyester 250 V dc working
C_4, C_5 330 μF 25 V dc working electrolytic

D1 TIL 209 red LED
TR1–TR4 BC107 (or similar) *npn* silicon

dc Multimeter

All resistors ±1% $\frac{1}{2}$ W metal oxide or thick film, high-stability types unless otherwise stated
R_1 180 Ω
R_2 18 Ω
R_3 2 Ω wire-wound (as close tolerance as possible)
R_4 43 kΩ + 47 kΩ + 8.2 kΩ connected in series
R_5 200 kΩ
R_6 430 kΩ + 270 kΩ connected in series
R_7 2 MΩ (use two 1 MΩ resistors connected in series if necessary)
R_8 82 kΩ ±5% $\frac{1}{2}$ W carbon film
VR_1 1 kΩ linear carbon preset potentiometer
VR_2 22 kΩ linear carbon potentiometer
100 μA meter movement
PP3 battery

High Input Impedance Voltmeter

All resistors ±1% $\frac{1}{2}$ W metal oxide or thick film, high-stability types unless otherwise state
R_1 9 MΩ (this value will be difficult to obtain: use a 9.1 MΩ or an 8.2 MΩ + 1.0 MΩ connected in series)
R_2 910 kΩ
R_3 91 kΩ
R_4 10 kΩ
R_5 1 MΩ
R_6 1 kΩ
R_7 56 kΩ
R_8 43 kΩ
R_9 100 kΩ
C_1 0.1 μF polyester 250 V dc working
IC1 TLO 81CP op-amp
D1, D2 IN914 Switching Diode
100 μA meter movement
S1 single-pole 4-position rotary switch

Simple Capacitance Bridge

R_1 10 kΩ ±10% $\frac{1}{2}$ W carbon film

R_4 10 kΩ linear carbon potentiometer with switch
C_1, C_2 10 nF polyester 250 V dc working
C_3 100 pF silvered mica 350 V dc working 1% tolerance
C_4 1 nF silvered mica 350 V dc working 1% tolerance
C_5 10 nF silvered mica 350 V dc working 1% tolerance
C_6 100 nF Mylar film 100 V dc working 10% tolerance
C_7 40 μF 16 V dc working electrolytic
TR1, TR2 BC107 (or similar) *npn* silicon
S1 2-pole 4-position rotary switch
X1 crystal earpiece

Electrolytic Capacitor Tester

R_1 220 Ω ±10% $\frac{1}{2}$ W carbon film
R_2 10 Ω ±10% $\frac{1}{2}$ W carbon film
VR_1 1 kΩ linear wire-wound 3 W potentiometer
S1 2-pole 3-position rotary switch

Transistor Tester

R_1 910 kΩ ±5% $\frac{1}{4}$ W carbon film
R_2, R_3 100 Ω ±10% $\frac{1}{4}$ W carbon film
R_4 2.2 kΩ ±5% $\frac{1}{4}$ W carbon film
D1 IN914 switching diode
S1 press-to-make push button
S2 2-pole 3-position rotary switch
S3 press-to-make push button
5 mA meter movement

Oscilloscope Calibrator

R_1, R_4, R_5 1 kΩ ±10% $\frac{1}{4}$ W carbon film
R_2, R_3 270 kΩ ±10% $\frac{1}{4}$ W carbon film
R_6 470 Ω ±10% $\frac{1}{4}$ W carbon film
R_7 330 Ω ±10% $\frac{1}{4}$W carbon film
R_8 1 kΩ ±1% thick film
R_9 100 Ω ±1% thick film
R_{10} 10 Ω ±1% thick film
R_{11} 1 Ω ±5% 2 W wire-wound
VR_1 10 kΩ linear carbon preset potentiometer
VR_2 100 Ω linear carbon preset potentiometer
C_1, C_2 47 nF polyester 250 V dc working
D1–D6 IN914 switching diode

Z1 BZX83 6.2 V Zener
TR1–TR3 BC107 (or similar) *npn* silicon
S1 S.P.D.T. miniature toggle switch

Oscilloscope Dual Trace Adaptor

All resistors ± 10% $\frac{1}{4}$ W carbon film unless otherwise stated
R_1, R_8 470 Ω
R_2, R_6 47 kΩ
R_3, R_5 22 kΩ
R_4, R_7 56 kΩ
$R_9, R_{12}, R_{13}, R_{14}$ 4.7 kΩ
R_{10}, R_{11}, R_{20} 82 kΩ
R_{15} 680 Ω
R_{16} 18 kΩ
R_{17}, R_{18} 8.2 kΩ
R_{19} 120 kΩ
R_{21} 5.6 kΩ
VR_1–VR_3 4.7 kΩ linear carbon potentiometers
C_1 50 μF 25 V dc working electrolytic
$C_2, C_7, C_{13}, C_{15}, C_{16}, C_{17}$ 12 μF 25 V dc working electrolytic
C_3, C_5 220 nF polyester 250 V dc working
C_4, C_6, C_{14} 10 μF 25 V dc working electrolytic
C_8, C_{10} 1 nF polyester 250 V dc working
C_9, C_{11} 5 nF polyester 250 V dc working
C_{12} 47 nF polyester 250 V dc working
D1–D5 IN914 switching diodes
TR1–TR6 2N3904 *npn* switching transistors
S1 S.P.S.T. miniature toggle switch
S2 D.P.D.T. miniature toggle switch

Simple Sawtooth Generator

All resistors ±10% $\frac{1}{4}$ W carbon film unless otherwise stated
R_1 27 kΩ
R_2 470 kΩ
R_3 390 Ω
R_4, R_5 220 Ω
R_6 4.7 kΩ
VR_1 47 kΩ linear carbon potentiometer
VR_2 4.7 kΩ linear carbon potentiometer

C_1 1 μF polyester 250 V dc working
C_2 10 μF 25 V dc working electrolytic
TR1, TR3 BC107 (or similar) *npn* silicon
TR2 2N4871 Unijunction Transistor

Low-noise Switch

All resistors ±10% $\frac{1}{4}$ W carbon film
R_1, R_2 33 kΩ
R_3 330 kΩ
C_1, C_2 100 nF polyester 250 V dc working
TR1 2N3819 *n* channel JUGFET
S1 S.P.D.T. miniature toggle switch

High Input Impedance Buffer Amplifier

All resistors ±10% $\frac{1}{4}$ W carbon film
R_1 1 MΩ
R_2 6.8 kΩ
R_3 1 kΩ
C_1 100 nF polyester 250 V dc working
C_2 47 μF 16 V dc working electrolytic
TR1 2N3819 *n* channel JUGFET
TR2 BC109 (or similar) low noise high gain *npn* silicon

A Bench Test Audio Amplifier

All resistors ±5% $\frac{1}{4}$ W carbon film unless otherwise stated
R_1 22 kΩ
R_2, R_3 100 kΩ
R_4, R_9 1.2 kΩ
R_5 100 Ω
R_6 2.7 kΩ
R_7 82 Ω
R_8 270 Ω
R_{10} 1.8 kΩ
R_{11}, R_{12} 1 Ω ±5% 3 W wire-wound
C_1 100 nF polyester 250 V dc working
C_2 4.7 μF 16 V dc working electrolytic
C_3, C_6 100 μF 16 V dc working electrolytic
C_4 5.6 nF polystyrene 160 V dc working

C_5 22 pF polystyrene 160 V dc working
C_7 470 μF 16 V dc working electrolytic
TR1 BC159 *pnp* a.f. small signal plastic case silicon
TR2, TR5 BC168 *npn* a.f. signal plastic case silicon
TR3 BC327 *npn* medium power (625 mW) silicon
TR4 BC337 *pnp* medium power (625 mW) silicon

r.f. Probe II

All resistors ±5% $\frac{1}{4}$ W carbon film
R_1 2.2 MΩ
R_2 5.6 kΩ
R_3 1 kΩ
C_1, C_2 10 nF polyester 160 V dc working
C_3, C_4 10 μF 16 V dc working electrolytic
TR1 2N3819 *n* channel JUGFET
D1 0A91 germanium point contact diode

Simple Timer Circuit

All resistors ±5% $\frac{1}{2}$ W carbon film
R_1 10 kΩ
R_2 1 MΩ
R_3 220 kΩ
R_4 100 Ω
VR_1 2 MΩ linear carbon potentiometer
C_1 220 μF 16 V dc working electrolytic
C_2, C_3 10 nF polyester 160 V dc working
TR1 BC157 *pnp* a.f. small signal plastic case silicon
IC1 NE 555 V Timer IC
S1 D.P.D.T. miniature toggle switch

Resistance Substitution Box

All resistors ±5% $\frac{1}{2}$ W carbon film
R_1 10 Ω
R_2 56 Ω
R_3 100 Ω
R_4 220 Ω
R_5 560 Ω
R_6 1 kΩ
R_7 2.2 kΩ
R_8 3.3 kΩ

R_9 5.6 kΩ
R_{10} 8.2 kΩ
R_{11} 10 kΩ
R_{12} 15 kΩ
R_{13} 22 kΩ
R_{14} 33 kΩ
R_{15} 56 kΩ
R_{16} 100 kΩ
R_{17} 220 kΩ
R_{18} 560 kΩ
R_{19} 820 kΩ
R_{20} 1 MΩ
R_{21} 5.6 MΩ
R_{22} 10 MΩ
S1, S2 single pole 12-position rotary switches
S3 S.P.D.T. miniature toggle switch

Capacitor Substitution Box

All working voltages are dc unless otherwise state
C_1 2.2 pF 350 V mica
C_2 10 pF 350 V mica
C_3 100 pF 350 V mica
C_4 560 pF 350 V mica
C_5 1 nF 500 V disc ceramic
C_6 2.2 nF 350 V mica
C_7 5.6 μF 160 V polystyrene
C_8 10 nF 350 V mica
C_9 22 nF 250 V polycarbonate
C_{10} 56 nF 250 V polycarbonate
C_{11} 100 nF 250 V polycarbonate
C_{12} 1 μF 500 V electrolytic
C_{13} 2.2 μF 500 V electrolytic
C_{14} 6.8 μF 500 V electrolytic
C_{15} 10 μF 500 V electrolytic
C_{16} 22 μF 500 V electrolytic
C_{17} 33 μF 63 V electrolytic
C_{18} 68 μF 63 V electrolytic
C_{19} 100 μF 63 V electrolytic
C_{20} 220 μF 63 V electrolytic
C_{21} 470 μF 63 V electrolytic
C_{22} 1000 μF 63 V electrolytic
S1, S2 single pole 12-position rotary switches
S3 S.P.D.T. miniature toggle switch

Appendix II: Component Suppliers

It is fortunate that there are very few 'custom-made' electronic components, and that most circuits can be constructed from parts that are easily available in most towns and cities. Indeed, circuit designers (particularly amateur ones) tend to take some trouble to avoid putting in parts that are unusual or difficult to get.

Most of the circuits and layouts in this book can be made with standard components, but inevitably there are a few items that may be hard to locate. A few suppliers are mentioned in the text, but I am providing a list here of those manufacturers and distributors that I have found helpful with 'difficult' parts.

Radiospares Ltd.
Radiospares can supply most components, via the trade. *Note that they will not supply direct to individuals, but only to shops and industrial users who have an account with them.* They are, however, fast and generally have everything in stock. Your local radio shop should be able to order components for you, and get them within a week. I have quoted 'RS' parts in many instances in the book, but I have also given specifications so that components for alternative suppliers could generally be used.

Maplin Electronic Supplies
(P.O. Box 3, Rayleigh, Essex SS6 8LR)
Supply over the counter or by mail order, in any quantity. They stock a very large range of components of all sorts, and publish a comprehensive catalogue which is well worth having.

Vero Electronics Ltd
(Industrial Estate, Chandlers Ford, Eastleigh, Hants)
Manufacturers of various high-quality printed-circuit boards, high-quality plastic cases, etc. They supply via the trade.

Semiconductor Supplies (Croydon) Ltd
(Orchard Works, Church Lane, Wallington, Surrey SM6 7NF)
Electronic components, by mail order only. They have unusually large stocks of ICs, transistors and diodes.

Marshalls
(40 Cricklewood Broadway, NW2 3ET and 325 Edgeware Road, W2)
Supply a wide range of small components both over the counter and by mail order. From time to time they have bargain offers which are well worth investigating.